Advances in 21st Century Human Settlements

Indexed by SCOPUS

This Series focuses on the entire spectrum of human settlements—from rural to urban, in different regions of the world, with questions such as: What factors cause and guide the process of change in human settlements from rural to urban in character, from hamlets and villages to towns, cities and megacities? Is this process different across time and space, how and why? Is there a future for rural life? Is it possible or not to have industrial development in rural settlements, and how? Why does 'urban shrinkage' occur? Are the rural areas urbanizing or is that urban areas are undergoing 'ruralisation' (in form of underserviced slums)? What are the challenges faced by 'mega urban regions', and how they can be/are being addressed? What drives economic dynamism in human settlements? Is the urban-based economic growth paradigm the only answer to the quest for sustainable development, or is there an urgent need to balance between economic growth on one hand and ecosystem restoration and conservation on the other—for the future sustainability of human habitats? How and what new technology is helping to achieve sustainable development in human settlements? What sort of changes in the current planning, management and governance of human settlements are needed to face the changing environment including the climate and increasing disaster risks? What is the uniqueness of the new 'socio-cultural spaces' that emerge in human settlements, and how they change over time? As rural settlements become urban, are the new 'urban spaces' resulting in the loss of rural life and 'socio-cultural spaces'? What is leading the preservation of rural 'socio-cultural spaces' within the urbanizing world, and how? What is the emerging nature of the rural-urban interface, and what factors influence it? What are the emerging perspectives that help understand the human-environment-culture complex through the study of human settlements and the related ecosystems, and how do they transform our understanding of cultural landscapes and 'waterscapes' in the 21st Century? What else is and/or likely to be new vis-à-vis human settlements—now and in the future? The Series, therefore, welcomes contributions with fresh cognitive perspectives to understand the new and emerging realities of the 21st Century human settlements. Such perspectives will include a multidisciplinary analysis, constituting of the demographic, spatio-economic, environmental, technological, and planning, management and governance lenses.

If you are interested in submitting a proposal for this series, please contact the Series Editor, or the Publishing Editor:

Bharat Dahiya (bharatdahiya@gmail.com) or

Loyola D'Silva (loyola.dsilva@springer.com)

More information about this series at http://www.springer.com/series/13196

Oscar Carracedo García-Villalba
Editor

Resilient Urban Regeneration in Informal Settlements in the Tropics

Upgrading Strategies in Asia and Latin America

Editor
Oscar Carracedo García-Villalba
Director Master of Urban Design,
Director Designing Resilience in Asia
International Research Programme,
Department of Architecture,
School of Design and Environment
National University of Singapore
Singapore, Singapore

ISSN 2198-2546 ISSN 2198-2554 (electronic)
Advances in 21st Century Human Settlements
ISBN 978-981-13-7309-1 ISBN 978-981-13-7307-7 (eBook)
https://doi.org/10.1007/978-981-13-7307-7

This Springer imprint is published by the registered company Springer Nature Singapore Pte Ltd.
The registered company address is: 152 Beach Road, #21-01/04 Gateway East, Singapore 189721, Singapore

Foreword

Towards Resilient Urban Regeneration of Informal Settlements in Asia and Latin America

The *Sustainable Development Goals Report 2019* underlined that one out of four urban residents in the world lived in slum-like conditions in 2018 (United Nations 2019; see Fig. 1). Highly relevant to public policy for cities, this piece of statistics was recorded despite the fact that by 2010 the lives of 227 million slum dwellers were improved under the Millennium Development Goal "slum target": "Achieve, by 2020, a significant improvement in the lives of at least 100 million slum dwellers" (UN-Habitat 2010; United Nations 2020a). Under the progress on Sustainable Development Goal 11, "to make cities and human settlements inclusive, safe, resilient and sustainable" (United Nations 2020), the "Sustainable Development Goals Report 2019" further noted that only half (53%) of urban residents had convenient access to public transport in 2018, and that two billion people did not have access to solid waste collection services (United Nations 2019). These are some of the major challenges that cities and towns around the world are confronted with at present.

Within this global urban context, *Resilient Urban Regeneration in Informal Settlements in the Tropics: Upgrading Strategies in Asia and Latin America*, edited by Oscar Carracedo García-Villalba, is a unique effort that brings together multiple perspectives to review and understand two geographically and historically distinctive urban contexts, tropical Asia and Latin America. These two continental regions at first sight might appear very different from their cultural, political, economic and social points of view. However, they feature certain common characteristics such as developing economies, rapid urban demographic growth, extreme poverty and deprivation, social exclusion, environmental degradation, impacts of climate change and a host of other sustainable urban development challenges (see Dahiya 2012a, b, 2014, 2016; Dahiya and Pugh 2000; Dahiya and Das 2020a, b; Das and Dahiya 2020; Haase et al. 2018; Asian Development Bank and Inter-American Development Bank 2014; UN-Habitat and ESCAP 2010, 2015; UN-Habitat 2012a, c; UN-ESCAP and UN-Habitat 2019).

MAKE CITIES AND HUMAN SETTLEMENTS INCLUSIVE, SAFE, RESILIENT AND SUSTAINABLE

1 OUT OF 4 URBAN RESIDENTS LIVE IN SLUM-LIKE CONDITIONS (2018)

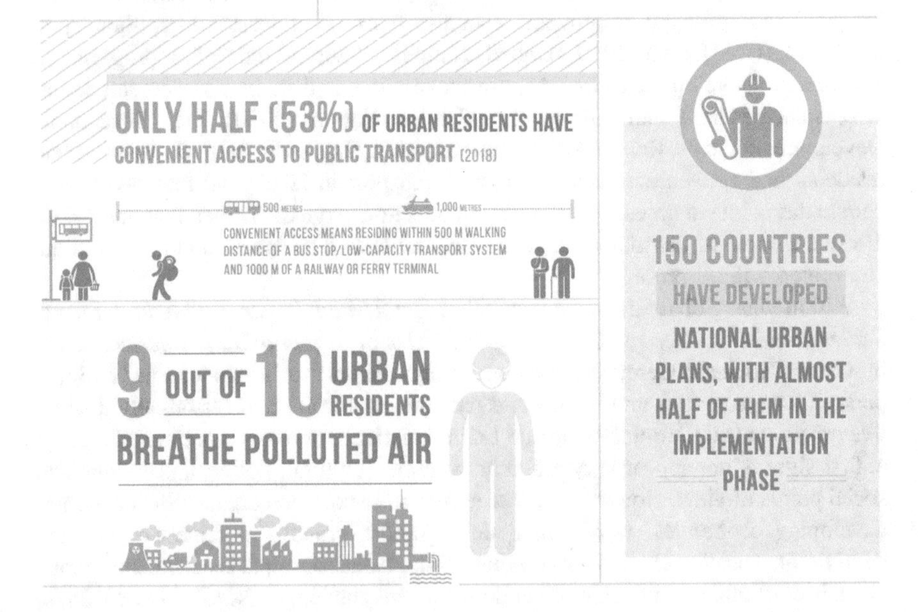

Fig. 1 Progress on Sustainable Development Goal 11 by 2018. *Source* United Nations (2019, p. 14)

Oscar Carracedo, an architect, urbanist and educator from the Catalonian city of Barcelona, takes his work on urbanism and architecture seriously. For more than two decades, he has been devoted to the task of bringing urbanism and architecture back to what he thinks is their primary purpose, serving people in the aspiration of creating a better built environment for a better society. This task brought him to focus his interest on what Cynthia E. Smith calls "design for the other 90%" (see Smith 2011; Smith and Smithsonian Institution 2011), and more specifically an interest in the physical, political, economic and social context of the almost one billion people, or 30% of the global urban population, that lives or survives in low-income, informal settlements—often called slums—in cities around the world. As a result of his professional interest in informal settlements, Oscar Carracedo started a modest project to develop a study and documentation of good practices—that can drive change and foster social solidarity—in transforming and improving the quality of life and the built environment of informal settlements.

The book is an exploration that can be contextualized vis-à-vis the intersections between contemporary and future urbanization processes, climate change, urban resilience, sustainable cities and the changing economies and societies. It is a contribution to the broader discussions about strategic approaches to on-site upgrading that has been ongoing for many years, since the writings by John F. C. Turner (see Turner 1976; Turner and Fichter 1972); the various reports and documents prepared by the United Nations' agencies (UN-Habitat and ESCAP 2010, 2015; UN-Habitat 1987, 1996, 1997, 2001, 2003a, b, 2004, 2006, 2007, 2008a, b, 2009, 2010a, b, 2011, 2012a, b, c, d, 2013a, b, 2014a, b, 2016, 2018; European Commission and UN-Habitat 2016; UN-ESCAP and UN-Habitat 2019; UN-ESCAP and UN-ECLAC, UN-HABITAT and Columbia University 2011; United Nations 1976, 2017; UNICEF and WHO 2019) and international development banks, including the African Development Bank, Asian Development Bank, the Inter-American Development Bank and the World Bank (see Nitti and Dahiya 2004; Bigio and Dahiya 2003, 2004; Linn 1983; Imparato and Ruster 2003; Gilbert 2004; Steinberg and Lindfield 2011; Bouskela et al. 2016; Gómez-Álvarez et al. 2017; Kaw et al. 2020; Asian Development Bank and Inter-American Development Bank 2014; African Development Bank 2011; World Bank Group 2014; African Development Bank, Asian Development Bank, European Bank for Reconstruction and Development and Inter-American Development Bank 2019); the globally adopted Millennium Development Goals and the subsequent Sustainable Development Goals (see United Nations 2015a, b, 2019, 2020); the "unslumming" approach by Jane Jacobs (Jacobs 1961); the "right to the city" argument by Lefebvre (1968, 1996); and the other notable reads such as "Planet of Slums" (Davis 2006) and "Arrival City" (Saunders 2012).

Thus, the book is an urgent call for action concerning the planning and governance of the urban places where over one billion people live, work and play. Moreover, this book brings fresh contributions on how to tackle the issues of urbanization and informality providing sensitive solutions. In doing so, it considers the shift towards on-site upgrading and regularization of informal settlements to resolve the challenges of improving access to basic infrastructure and the formalization of land tenure, in order to improve the resilience and quality of life of urban dwellers in the two developing regions of Asia and Latin America.

This collection of essays is also an important addition to the existing knowledge on the on-site slum-upgrading policies, approaches and strategies implemented in urban informal settlements in Asia (see Boonyabancha 2009; Boonyabancha and Kerr 2018) and Latin America by finding similarities between the two cases. The proposed strategic approaches through spatial connections, social cohesion, capacity building and institutional coordination constitute a good base that can help in the formulation of adequate resilient planning for low-income settlements. This book is a useful resource and a guide for architects, planners, policymakers, built environment professionals, development practitioners, humanitarian and development aid agencies, non-governmental organisations, and many other stakeholders that are involved in the regeneration of urban informal settlements.

All in all, this edited volume is an excellent contribution to the debates on on-site upgrading and regeneration strategies for informal settlements and alternative systems to provide access to housing for all urban residents. Through the case studies, the contributing authors underline specific urban actions focused on the on-site upgrading or regeneration of informal settlements as a way to improve the living conditions in these areas and building more equitable and resilient cities.

Bharat Dahiya
Director
Research Center for Integrated
Sustainable Development
College of Interdisciplinary Studies
Thammasat University
Bangkok, Thailand

References

African Development Bank (2011) Transforming Africa's cities and towns into engines of economic growth and social development: the bank group's urban development strategy. African Development Bank, Operational Resources and Policies Department, Abidjan

African Development Bank, Asian Development Bank, European Bank for Reconstruction and Development & Inter-American Development Bank (2019) Creating livable cities: regional perspectives. Asian Development Bank, Mandaluyong City, Philippines

Asian Development Bank & Inter-American Development Bank (2014) Sustainable urbanization in Asia and Latin America. Asian Development Bank, Mandaluyong City, Philippines

Bigio AG, Dahiya B (2003) World bank investments for the urban environment, environment strategy notes no. 8. Report No. 53364, The World Bank, Washington, DC. http://documents.worldbank.org/curated/en/366001468151475459/World-Bank-investments-for-the-urban-environment

Bigio AG, Dahiya B (2004) Urban environment and infrastructure: toward livable cities. Directions in development series. The World Bank, Washington DC. https://openknowledge.worldbank.org/handle/10986/15018 License: CC BY 3.0 IGO

Boonyabancha S (2009) Land for housing the poor—by the poor: experiences from the Baan Mankong nationwide slum upgrading programme in Thailand. Environ Urban 21(2):309–329. https://doi.org/10.1177/0956247809342180

Boonyabancha S, Kerr T (2018) Lessons from CODI on co-production. Environ Urban 30(2):444–460. https://doi.org/10.1177/0956247818791239

Bouskela M, Casseb M, Bassi S, De Luca C, Facchina M (2016) The road toward smart cities: migrating from traditional city management to the smart city. Inter-American Development Bank (IDB), Washington DC

Dahiya B, Pugh C (2000) The localisation of agenda 21 and the sustainable cities programme. In: Pugh C (ed) Sustainable cities in developing countries: theory and practice at the millennium, pp 152–184, Earthscan, London

Dahiya B (2012a) 21st century Asian cities: unique transformation, unprecedented challenges. Glob Asia 7(1):96–104 (ISSN: 1976-068X)

Dahiya B (2012b) Cities in Asia, 2012: demographics, economics, poverty, environment and governance. Cities, 29(Supplement 2):S44–S61. Retrieved from https://doi.org/10.1016/j.cities.2012.06.013

Dahiya B (2014) Southeast Asia and sustainable urbanization. Glob Asia 9(3):84–91 (ISSN: 1976-068X)

Dahiya B (2016) ASEAN economic integration and sustainable urbanization. J Urban Culture Res 12:8–14. https://doi.org/10.14456/jucr.2016.10

Dahiya B, Das A (eds) (2020a) New urban agenda in Asia-Pacific: governance for sustainable and inclusive cities. Springer, Singapore. https://doi.org/10.1007/978-981-13-6709-0

Dahiya B, Das A (2020b) New urban agenda in Asia-Pacific: governance for sustainable and inclusive cities. In: Dahiya B, Das A (eds) New urban agenda in Asia-Pacific. In: Advances in 21st century human settlements. Springer, Singapore, pp 3–36. https://doi.org/10.1007/978-981-13-6709-0_1

Das A, Dahiya B (2020) Towards inclusive urban governance and planning: emerging trends and future trajectories. In: Dahiya B, Das A (eds) New urban agenda in Asia-Pacific. Advances in 21st century human settlements. Springer, Singapore, pp 353–384. https://doi.org/10.1007/978-981-13-6709-0_13

Davis M (2006) Planet of slums. Verso, New York

European Commission & UN-Habitat (2016) The state of European Cities 2016: cities leading the way to a better future. Publications Office of the European Union, Luxembourg

Gilbert R (2004) Improving the lives of the poor through investment in cities: an update on the performance of the World Bank's urban portfolio. World Bank Operations Evaluation Department, Washington DC

Gómez-Álvarez D, Rajack R, López-Moreno E, Lanfranchi G (eds) (2017) Steering the metropolis: metropolitan governance for sustainable urban development. Inter-American Development Bank (IDB), Washington DC

Haase D, Guneralp B, Dahiya B, Bai X, Elmqvist T (2018) Global urbanization: perspectives and trends. In: Elmqvist T, Bai X, Frantzeskaki N, Griffith C, Maddox D, McPhearson T, Parnell S, Romero-Lankao P, Simon D, Watkins M (eds) Urban planet: knowledge towards sustainable cities. Cambridge University Press, Cambridge, pp 19–44. https://doi.org/10.1017/9781316647554.003

Imparato I, Ruster J (2003) Slum upgrading and participation: lessons from Latin America. Directions in development series. The World Bank, Washington DC. https://openknowledge.worldbank.org/handle/10986/15133 License: CC BY 3.0 IGO

Jacobs J (1961) The death and life of Great American Cities. Vintage Books, New York

Kaw JK, Lee H, Wahba S (2020) The hidden wealth of cities: creating, financing, and managing public spaces. The World Bank, Washington, DC. https://openknowledge.worldbank.org/handle/10986/33186 License: CC BY 3.0 IGO

Lefebvre H (1968) Le droit à la ville. Anthropos, Paris

Lefebvre H (1996) Writings on cities (trans: Kofman E, Lebas E). Blackwell, Cambridge, MA

Linn JF (1983) Cities in the developing World: policies for their equitable and efficient growth. Oxford University Press, New York

Nitti R, Dahiya B (2004) Community driven development in urban upgrading, social development notes, no. 85, report no. 29926. The World Bank, Washington DC. http://documents.worldbank.org/curated/en/245161468779979383/Community-driven-development-in-urban-upgrading

Saunders D (2012) Arrival city: how the largest migration in history is reshaping our world. Random House, New York

Smith CE (2011) Design with the other 90%: cities. Places J. https://doi.org/10.22269/111017. Accessed 20 Apr 2020

Smith CE, Smithsonian Institution (2011) Design with the other 90%: cities. Cooper Hewitt National Design Museum, New York (ISBN: 9780910503839)

Steinberg F, Lindfield M (eds) (2011) Inclusive cities. Urban development series. Asian Development Bank, Mandaluyong City, Philippines

Turner JFC (1976) Housing by people: towards autonomy in building environments. Marion Boyars, London

Turner JFC, Fichter R (Eds) (1972) Freedom to build: dweller control of the housing process. Macmillan, New York

United Nations (1976) The Vancouver declaration on human settlements. 31 May to 11 June. United Nations Conference on Human Settlements, Vancouver. https://mirror.unhabitat.org/downloads/docs/The_Vancouver_Declaration.pdf. Accessed 5 May 2020

United Nations (2015a) Millennium development goals report 2015. United Nations, New York

United Nations (2015b) Transforming our world: the 2030 agenda for sustainable development, A/RES/70/1. https://sustainabledevelopment.un.org/post2015/transformingourworld/publication. Accessed 20 Apr 2020

United Nations (2017) The new urban agenda, A/RES/71/256, Habitat III and United Nations

United Nations (2019) The sustainable development goals report 2019. United Nations, New York. https://unstats.un.org/sdgs/report/2019/. Accessed 4 May 2020

United Nations (2020a) United Nations millennium development goals: goal 7: ensure environmental sustainability. https://www.un.org/millenniumgoals/environ.shtml. Accessed 4 May 2020

United Nations (2020b) Sustainable development goal 11: make cities and human settlements inclusive, safe, resilient and sustainable. https://sustainabledevelopment.un.org/sdg11. Accessed 20 Apr 2020

UN-ESCAP and UN-Habitat (2019) The future of Asian and Pacific Cities 2019: transformative pathways towards sustainable urban development. UN-ESCAP, Bangkok

UN-Habitat & UN-ESCAP (2010) The state of Asian cities 2010/11. UN-Habitat, Fukuoka

UN-Habitat & UN-ESCAP (2015) The state of Asian and Pacific cities 2015: Urban transformations shifting from quantity to quality. UN-Habitat and ESCAP, Nairobi and Bangkok

UN-Habitat (1987) Global report on human settlements 1986. Oxford University Press, Oxford

UN-Habitat (1996) An urbanizing world: global report on human settlements, 1996. Oxford University Press, Oxford

UN-Habitat (1997) The Istanbul declaration and the habitat agenda. UN-Habitat, Nairobi

UN-Habitat (2001) The state of the world's cities. UN-Habitat, Nairobi

UN-Habitat (2003a) The challenge of slums: global report on human settlements 2003. Earthscan, London

UN-Habitat (2003b) Improving the lives of 100 million slum dwellers: guide to monitoring target 11—progress towards the millennium development goals. UN-Habitat Global Urban Observatory, Nairobi

UN-Habitat (2004) The state of the world's cities 2004/2005: globalization and urban culture. Earthscan, London

UN-Habitat (2006) State of the world's cities 2006/7: the millennium development goals and urban sustainability—30 years of shaping the habitat agenda. UN-Habitat, Nairobi

UN-Habitat (2007) Enhancing urban safety and security: global report on human settlements 2007. Earthscan, London

UN-Habitat (2008a) State of the world's cities 2008/2009: harmonious cities. UN-Habitat, Nairobi

UN-Habitat (2008b) The state of African Cities 2008: a framework for addressing urban challenges in Africa. UN-Habitat, Nairobi

UN-Habitat (2009) Planning sustainable cities: global report on human settlements 2009. Earthscan, London

UN-Habitat (2010a) State of the world's cities 2010/2011: bridging the urban divide. Earthscan, London

UN-Habitat (2010b) The state of African Cities 2010: governance, inequality and urban land markets. UN-Habitat, Nairobi

UN-Habitat (2011) Cities and climate change: global report on human settlements 2011. Earthscan, London

UN-Habitat (2012a) The State of Latin American and Caribbean cities 2012: towards a new urban transition. UN-Habitat, Nairobi

UN-Habitat (2012b) State of the world's cities 2012/2013: prosperity of cities. UN-Habitat, Nairobi

UN-Habitat (2012c) Sustainable urbanization in Asia: a sourcebook for local governments. UN-Habitat, Nairobi

UN-Habitat (2012d) The state of Arab Cities 2012: challenges of urban transition. UN-Habitat, Nairobi

UN-Habitat (2013a) Planning and design for sustainable urban mobility: global report on human settlements 2013. Routledge, New York

UN-Habitat (2013b) The State of European cities in transition 2013: taking stock after 20 years of reform. UN-Habitat, Nairobi

UN-Habitat (2014a) The State of African cities 2014: re-imagining sustainable urban transitions. UN-Habitat, Nairobi

UN-Habitat (2014b) Practical guide to designing, planning and implementing citywide slum upgrading programs. UN-Habitat, Nairobi

UN-Habitat (2016) World cities report 2016—urbanization and development: emerging futures. UN-Habitat, Nairobi

UN-Habitat (2018) The State of African Cities 2018: the geography of African investment. UN-Habitat, Nairobi

UN-ESCAP and UN-ECLAC, UN HABITAT & Columbia University (2011) Are we building competitive and liveable cities?: Guidelines for developing eco-efficient and socially inclusive infrastructure. Bangkok: UN-ESCAP

UNICEF (United Nations Children's Fund) & WHO (World Health Organization) (2019) Progress on household drinking water, sanitation and hygiene 2000–2017: special focus on inequalities. UNICEF and WHO, New York

World Bank Group (2014) The Asian coalition for community action's approach to slum upgrading. The World Bank, Washington, DC. https://openknowledge.worldbank.org/handle/10986/20100 License: CC BY 3.0 IGO

Preface

Informal settlements and low-income areas are one of the main expressions and consequences of urbanization in our twenty-first-century cities. According to the United Nations estimates, 863 million people—a quarter of the world's urban population—live in slums and under informal conditions, and this population is likely to reach a total of 1 billion by 2020 (UN-Habitat 2013). It is interesting to note that most of this informal population, a total of 55%, is located in the tropics region, between the tropics of Cancer and Capricorn. This book focuses on the implementation of slum-upgrading projects in the tropical region and contrasts socially resilient regeneration experiences in Latin America and Asia. Although informality in Latin America is in its second generation compared to Asia, the book aims to find points of connection and similarities between both cases. The book follows some of the strategies proposed by UN-Habitat/CAF such as the urban action framework against inequality (Vaggione 2014). This book aims to learn from the Latin American and Asian experiences that have been the pioneers of some of the most successful urban regeneration projects in the world.

Singapore
2019

Oscar Carracedo García-Villalba

Reference

Vaggione P (2014) Construction of more equitable cities: public policies for inclusion in Latin America. Corporación Andina de Fomento. UN-HABITAT

Acknowledgements

This book is the result of the thinking process conducted through the research project "Spatial patterns and land tenure. Resilient urban planning for the informal city" (R-295-000-110-133), funded by the Ministry of Education (MOE) of Singapore, Academic Research Fund (AcRF), and some of the contents are partly based on the outcomes. As the principal investigator of this research project, I would like to thank the Ministry of Education and the National University of Singapore for their support in funding this research.

Also, my most sincere gratitude to Prof. Wong Yun Chii and Prof. Heng Chye Kiang from the Department of Architecture, School of Design and Environment at NUS, for their inspiration and continuous support since I joined the National University of Singapore in 2013.

My special and most sincere and deepest gratitude to Francesco Rossini, Reena Tiwari, Claudio Acioly, Clara Irazábal, David Gouverneur and Armando Arteaga for their enthusiastic support and their contributions to the book coming from many places across the world. Without their thoughts, this book would not have been possible.

Special thanks are due to Dr. Bharat Bahiya for his commitment to addressing this topic as the editor of this book series and also for his support during the writing process and his time and generous words in writing the foreword. Also, my gratitude to Springer and Loyola D'Silva for allowing me to contribute to the debates of human settlements in the current century.

My gratitude also goes to my research assistant Chaitali Dighe for her time, help and discussions during the research process and the book's edition.

My heartfelt thanks go to my friends, family, brothers, and especially to my mother and father for always being supportive in all my initiatives. To Unwin, my black lab, who unconditionally and tirelessly accompanied me around the world and during the first years of my research. And last but certainly not least, to Maite, for her love, encouragement and patience, and to my beloved daughter Maya, who has changed my life for the better, giving me the necessary energy to finish this endeavour.

Contents

Editor and Contributors

About the Editor

Oscar Carracedo García-Villalba is an Architect, Urbanist and Educator, currently Assistant Professor at the Department of Architecture, National University of Singapore. He is the director of the Master of Urban Design at NUS and director of the DRIA-Designing Resilience in Asia International Research Programme, where he develops his research on urban resilience, climate change, sustainability, integrated urban planning and informal urbanism practices and processes. His work explores the intersections between contemporary and future urbanization processes, climate change and sustainable futures, hybrid dense cities and urban resilience, and changing economies and societies.

Oscar is the author of numerous books and articles, and drawing on his rigorous research, he has recently published "Designing Resilience in Asia. Planning the unpredictable, designing with uncertainty" (ACTAR, 2020), "Silicon Singapore. Urban Projects for Hybrid and Resilient Innovation Districts" (Basheer, 2020), "Ibid./In the same place. Nine Lessons and Six Possibilities about On-site Resilient Revitalization Strategies for Informal Neighbourhoods" (ORO, 2016), 'Indus_hoods. From industries to Neighbourhoods' (CASA-JTC-i3, 2016), and 'Naturban. Barcelona's Natural Park, A rediscovered relation' (CoAC, 2015), among other books. He has also co-edited the book 'Advanced Studies in Energy Efficiency and Built Environment for Developing Countries' (Springer, 2019).

Oscar has been invited to present as a keynote speaker in 14 countries worldwide including South Africa, Malaysia, Germany, Korea, Thailand, Indonesia, Singapore, Taiwan, Japan, Hong Kong, the Philippines, Ecuador, Colombia, and Spain. He has also lectured at many universities such as Pratt Institute (USA), Stevens Institute of Technology (USA), IUA di Venezia (Italy), TU Darmstadt (Germany), NCKU (Taiwan), PUCE (Ecuador), EAFIT Medellín (Colombia), King Mongkut's University of Technology Thonburi (Thailand), Faculty of Architecture of the University of Ljubljana (Slovenia), and Kyushu University (Japan) among others. He is also actively engaged as an expert with UN-Habitat, World Bank and many other reputable international institutions.

Oscar is also the CEO of CSArchitects, an urban planning, urban design and architecture firm based in Barcelona, Spain. Spanning over 20 years of international professional experience, he has been responsible for more than 60 masterplans and urban-scale commissions, an extensive number of projects and consultancies in urban design, site, physical and spatial planning, over a dozen architecture and public space projects, as well as many projects with underprivileged communities. Oscar has won two national urban planning and design prizes, and his work and research has been awarded in more than 40 national and international competitions and published nationally and internationally.

Contributors

Claudio Acioly is an architect, urban planner and development practitioner with more than 35 years of experience in more than 30 countries working with local and national governments, NGO's, universities and international development organizations. After 15 years working internationally with the Rotterdam-based Institute for Housing and Urban Development Studies (IHS), Acioly worked as senior manager and housing expert with the United Nations Human Settlements Programme (UN-Habitat) during the period 2008–2019. He was the head of Housing Policy and coordinator of the UN Housing Rights programe, coordinator of the Advisory Group on Forced Evictions to the Executive Director of UN-Habitat—AGFE and head of Capacity Building and Training. He led the housing policy work of UN-Habitat in countries such as Cuba, Ghana, Malawi, El-Salvador, Uganda, Vietnam, Nepal and Ecuador and was directly involved in housing policy planning and implementation as well as slum upgrading. He is the lead author of key UN-Habitat strategies such as the street-led slum upgrading strategy, the housing profile methodology, the housing barometer and the housing rights index. Acioly led several global programs and initiatives linking capacity building, institutional development and policy change in various places in the world. As senior manager with the German Development Cooperation Agency (GIZ) Acioly was in 2020 the director of the European Union's International Urban Cooperation Programme for Latin America and the Caribbean.

Armando Arteaga is an architect, Master in Urban and Regional Studies (National University of Colombia), Master's Degree in Urbanism (Polytechnic University of Catalonia) and Ph.D. in Urbanism (Polytechnic University of Catalonia). Armando is currently a full time assistant professor at the School of Urban-Regional Planning of the Architecture Faculty, Universidad Nacional de Colombia—Medellín campus.

With awards in several architectural competitions and professional practice that includes public administration, private enterprise, consulting and project design, experience that contributes to a holistic vision of architecture and urban issues.

Published articles in magazines and books include: The transformation of public space; between theory, legislation and practice (2007); Public space: a conceptual approach (2018); Co-author of the book: Urbanism in Medellín, 21st century; Contributions to the debate (2018).

Research projects address public space in various facets: contemporary civic senses, everyday use, performance evaluation, project design logics, mobility, infrastructural spaces and the role of urban facilities in the city's reconfiguration.

David Gouverneur Received his M.Arch in Urban Design from Harvard University (1980) and his B.Arch from the Universidad Simón Bolívar in Caracas, Venezuela (1977). He was Chair of the School of Architecture at Universidad Simón Bolívar (1987–91), as well as a professor in this School's Departments of Architecture and City and Regional Planning from 1980 to 2008. From 1991 to 1994, he was the Director, and from 1995 to 1996, the Adjunct Secretary of Urban Development of Venezuela. He was cofounder and professor of the Urban Design program and Director of the Mayor's Institute in Urban Design at Universidad Metropolitana, both created with the support of Harvard University, in Caracas, Venezuela (1996–2008). He has taught at the University of Pennsylvania since 2002, first as a visiting lecturer in the Department of Landscape Architecture from 2002 to 2010, and in the Department of City and Regional Planning from 2009 to 2010. He has been an Associate Professor in the Department of Landscape Architecture since 2010, and since 2012 he is Professor in Practice of this Department.

His current area of research focuses on the notion of Informal Armatures, an alternative method to address the rampant urbanization in developing countries where a high percentage of the population already lives, and will live, in self-constructed areas. In light of the limited success of conventional planning, urban, design and housing policies, Informal Armatures may prove to be a powerful tool to foster the sustainable growth of informal settlements, as the dominant form of territorial occupation in the developing world. The goal is to allow them to perform a la par and surpass the vitality, economic drive and environmental qualities of formal urbanization. The ideas are condensed in his most recent publication, Planning and Design for New Informal Settlements: Shaping the Self-Constructed city and its updated version in Spanish *Diseño de nuevos asentamientos informales*.

Clara Irazábal is the Director of the Latinx and Latin American Studies Program and Professor of Planning in the Department of Architecture, Urban Planning and Design (AUPD) at the University of Missouri, Kansas City (UMKC). Before joining UMKC in 2016, she was the Latin Lab Director and Associate Professor of Urban Planning in the Graduate School of Architecture, Planning and Preservation at Columbia University, New York City. She got her Ph.D. from the University of California at Berkeley and has two master degrees. In her research and teaching, she explores the interactions of culture, politics, and placemaking, and their impact on community development and socio-spatial justice in Latin American cities and US Latinx and immigrant communities. Irazábal has published academic work in

English, Spanish, Portuguese, and Italian. She is the author of Urban Governance and City Making in the Americas: Curitiba and Portland (Ashgate, 2005) and the editor of Transbordering Latin Americas: Liminal Places, Cultures, and Powers (T) Here (Routledge 2014) and Ordinary Places, Extraordinary Events: Citizenship, Democracy, and Public Space in Latin America (Routledge 2008, 2015). Irazábal has worked as consultant, researcher, and/or professor in multiple countries of the Americas, Europe, and Asia.

Francesco Rossini is an architect and urban designer and is an Assistant Professor at the School of Architecture at the Chinese University of Hong Kong (CUHK). Since 2019, he is also an Adjunct Assistant Professor in Columbia GSAPP's Urban Planning program. At CUHK, Francesco contributes to teaching in both the Undergraduate and Master programs, and over the years he has developed a comprehensive approach where research, teaching, and design are mutually related and interconnected. Francesco graduated with honors from the faculty of Architecture of University of Napoli Federico II obtaining a master's degree in architecture. In 2014, he completed (cum laude) his Ph.D. in Urbanism from Polytechnic University of Catalonia (Barcelona Tech UPC), where he explored the role of public spaces in private developments in Hong Kong. During his doctoral studies, he was awarded a grant by the Ministry of Education of the Government of Spain. Before joining CUHK, Francesco Rossini has participated in different research programs, collaborating with the University of Napoli Federico II, Politecnico di Milano, Barcelona Tech UPC, and Tongji University in Shanghai. His work has given rise to a series of publications, journal articles, book chapters, conference presentations, installations, and international exhibitions. In the last five years, he realized different placemaking interventions in Hong Kong, Singapore, Kuala Lumpur, and Manila exploring how temporary urban actions influence the behavior of the people by provoking new social interactions.

Reena Tiwari is a Professor at Curtin University in Australia and a scholar in the field of sustainable community development. Her work with marginalised communities in the local, national and international context has been recognised and published as books and articles.

She has published extensively on space psychology and place-making; sustainable urbanism, and has forwarded a model of community engagement which is ethnographic, collaborative, and transdisciplinary, and has the goal of facilitating change and adaptation for all involved in the process.

Reena has worked with urban and rural based communities in India since 2004 around sustainable development projects funded by Aus-Aid, National Colombo Plan, and the Australia-India Council. These include Lakhnu rural development project, Slum Action project in Ahmedabad, Safe City Bengaluru related to walkability issues, and To Walk or Not to Walk project in New Delhi.

Reena has worked with Australian Aboriginal communities on development and design projects since 2015. These include the Wakuthuni Aboriginal community sustainable livelihood project in Pilbara; projects with the Stolen Generations Survivors and Bringing Them Home WA for redevelopment of Carrolup-Marribank and Wandering mission sites as healing centres; and the most recently funded project on Digital storytelling for Mogumber mission.

She has been a Visiting Professor and Scholar at the University of California Berkeley, University of Texas A & M and University Pierre Mendes in Grenoble and an adjunct Professor at the University International Catalunya in Barcelona.

Neeti Trivedi is currently a guest faculty for Urban Planning and Architectural Design at BNCA in Pune, India. She received her doctorate in Urban and Regional Planning from Curtin University, Western Australia in 2017, where she examined opportunities for capacity building of the urban poor through redevelopment interventions in their built environment. Her research interests lies in the challenges posed by the interplay of community participation and collaborative planning strategies adopted for community development.

Jessica Winters Jessica holds an honours Bachelor of Arts (Urban and Regional Planning) from Curtin University and Diploma in Project Management. She has managed several major housing and cultural development projects in metropolitan and regional Western Australia, and internationally.

Jessica is the former Director of Samaky Foundation, a grass-roots community development non-for-profit organisation. This organisation delivered education, capacity building, sanitation and land tenure programs in a small community on the urban fringe of Siem Reap, Cambodia.

Domestically, Jessica has been an active project manager, involved in projects from small community housing developments to a multi-million dollar cultural precinct development in the South West of Western Australia.

Concurrently, Jessica has been involved in a number of academic research projects including the Lakhnu Project, a research and community development partnership between Curtin University and NGO Indian Rural Education and Development (IREAD), including the co-authoring of journal article "The death of strategic plan: questioning the role of strategic plan in self-initiated projects relying on stakeholder collaboration".

Weaving her experience in community capacity building, project management and research, her goal in any project is to forge long-term community interaction that informs better decision making in government and empowers communities to make sustainable changes in their own environments.

Abbreviations

1SF	The One Safe Future Program
ABCD	Asset-based community development
ACCA	Asian Coalition for Community Action
ACHR	Asian Coalition of Housing Rights
AKAA	Aga Khan Award for Architecture
AI	Appreciative inquiry
ANGOC	Asian NGO Coalition for Agrarian Reform and Rural Development
APPUC	Assessoria de Pesquisa e Planejamento Urbano de Curitiba
BAPPE-MPMHT	Badon Pelaksana Pembangunan–Proyek Mohammad Hum Thamrin
BEA	Baan Eua Arthorn
BIP	BaSECo Implementation Plan
BMA	Bangkok Metropolitan Administration
BMP	Bann Mankong Programme
BMR	Bangkok Metropolitan Region
BSUP	Basic Services for the Urban Poor
CDHU	Housing and Urban Development Company
CINVIT	Centre for Research on Informality of the Universidad de Valparaiso
CKIP	Comprehensive Kampung improvement program
CMR	Curitiba's metropolitan region
CODI	Community Organisations Development Institute
COSI	Central Office for Slum Improvement
COMEC	Coordenação da Região Metropolitana de Curitiba (Curitiba's Metropolitan Region Coordination)
CONAVI	Venezuelan Housing Council
CPB	Crown Property Bureau
CPM	Consejería Presidencial para Medellín y su Área Metropolitana (Presidential Council for Medellín and its Metropolitan Area)

DILG	Department of Interior and Local Government
DNA	Deoxyribonucleic Acid
DS	Dubai Site
EAFIT	Escuela de Administración, Finanzas e Instituto Tecnológico (School of Administration, Finance and Technological Institute)
EDU	Empresa de Desarrollo Urbano (Urban Development Company)
GDP	Gross domestic product
GE	General electric
GHB	Government Housing Bank
GIS	Geographic Information Systems
GK	Gawad Kalinga
GPN	Global Planners' Network
HDI	Human Development Indexes
IA	Informal Armature
ICP	Innovation Centre for Poor
ICPP	Innovation Centre for Poor Project
IDB	Inter-American Development Bank
IFAD	International Fund for Agricultural Development
IIED	International Institute for Economic Development
IIRR	International Institution of Rural Reconstruction
IMF	International Monetary Fund
INR	Indian Rupee
IPPUC	Institute of Urban Research and Planning of Curitiba (Instituto de Pesquisa e Planejamento Urbano de Curitiba)
ISF	Informal settler families
ITS	Faculty of Architecture at the Institute of Technology Sepuluh
JNNURM	Jawaharlal Nehru National Urban Renewal Mission
KIP	Kampung Improvement Programme
LAC	Latin American and Caribbean
LPP	Laboratory for Housing and Human Settlements
MCE	Mapping and Charting Establishment (check)
MDG	Millennium Development Goals
MHT	Mahila Housing Trust
MHT-KIP	Mohammed Husni Thamrin Programme
MIDEPLAN	Ministerio de Planificación Nacional y Política Económica (Ministry of National Planning and Economic Policy)
MPPPD	Ministerio del Poder Popular para la Planificación y el Desarrollo (Ministry of People's Power for Planning and Development)
MoHUPA	Ministry of Housing and Urban Poverty Alleviation
NAFTA	North American Free Trade Agreement
NESDP	National Economic and Social Development Plans
NASSCO	National Shipyard and Steel Corporation

NGO	Non-Governmental Organization
NHA	National Housing Authority
NS	New Site
NUDP	National Urban Development Policy
NVC	Núcleo de Vida Ciudadana (Hubs of Civic Life)
PAC	Growth Acceleration Programme
PARC	Palestinian Agricultural Relief Committee
PAT	Port Authority of Thailand
PDDUA	Plano Diretor de Desenvolvimento Urbano e Ambiental de Porto Alegre (Master Plan of Urban Development and Environment of Porto Alegre)
PMC	Pune Municipal Corporation
PND	Plan Nacional de Desarrollo (National Development Plan)
POT	Plan de Ordenamiento Territorial (Plan of Territorial Ordering)
POUSO	Posto de Orientação Urbanistica e Social
PPS	Project for Public Space
PRA	Participatory rural appraisal
PRIMED	The Programa Integral de Mejoramiento de Barrios Subnormales en Medellín (Integral Substandard Neighbourhood Improvement Program in Medellín)
PUI	Proyecto Urbano Integral (Integral Urban Projects)
RTPI	Royal Town Planning Institute
SDG	Sustainable Development Goals
SEAM	Secretaria Extraordinária de Assuntos Metropolitanos (Special Secretary of Metropolitan Affairs)
SEHAB	City of São Paulo Secretariat of Housing
SELCO	Solar Electric Light Company
SEWA	Self-Employed Women's Association
SJDC	St John's Development Company
SMGP	Municipal System of Management of Planning
SRT	State Railways of Thailand
TAO	Technical Assistance and Organization
THB	Thai baht
UCDO	Urban Community Development Office
UDC	Urban Development Corporation
UDU	Urban Design Units
UDeCOTT	Urban Development Corporation of Trinidad and Tobago
UK	United Kingdom
UNCHS	United Nations Conference on Human Settlements
UN-Habitat	United Nations Human Settlements Programme
UPA	Urban Poor Associates
UPP	Unidade de Polícia (Pacifying Police Unit)
UPU	Urban Planning Units
URBAM	The Centre for Urban and Environmental Studies and professional practice

US	United States
USD	United States Dollar
WRS-KIP	Wage Rudolf Supratman Programme
YSUP	Yerwada Slum-Upgrading Project

List of Figures

Medellín, Urban Renewal of Informal Settlements Through Public Space: The Case of the North-Eastern Integral Urban Project (PUI)

Capacity Building and Education in Latin America. Contributions and Limitations of the Venezuelan and Colombian Experiences, Lessons Looking into the Future

List of Tables

Introduction, Findings, and Reflections

Oscar Carracedo García-Villalba

The formation of slums and informal settlements is one of the most visible manifestations of the world's current rapid urbanisation trends. In the 2016 Global Cities Report, UN-Habitat estimated that 32.7% of the world's urban population, around 1 billion people, was living in informal urban settlements. With business as usual and fast urbanisation, especially in countries in the Global South, it is expected to grow over one more billion by 2030. The report indicates that the proportion of the world's urban population living in informal settlements has decreased in the last two decades from 46.2% in 1990 to 32.6% in 2010 and 29.7% in 2014. However, in absolute numbers, the total population living in informal settlements is still growing, and in 2014 881 million people lived in informal settlements, compared to 791 million in 2000, and 689 million in 1990, which represents an increase of 28%, partly due to accelerating urbanisation, population growth and the lack of appropriate land and housing policies (UN-Habitat 2016).

Informal settlements are not just a result of population explosion, demographic change and rural-urban migration processes; they exist and keep growing further because of ineffective urban planning and regulatory systems, a failure of national and government housing policies, laws and delivery systems to meet demand, low investment in infrastructure, and the limited options for people with fewer economic resources to access the formal land and housing market.

In the Policy Focus Report (2011), Fernandes explains that urban planning tradition has always excluded urban informality and the urban poor, which, as a result, has reinforced the informal processes. The recurrent poor integration of land, housing,

O. Carracedo García-Villalba (✉)
Director Master of Urban Design, Director Designing Resilience in Asia International Research Programme, Department of Architecture, School of Design and Environment, National University of Singapore, Singapore, Singapore
e-mail: oscar_carracedo@nus.edu.sg; omc@coac.net

O. Carracedo García-Villalba (ed.), *Resilient Urban Regeneration in Informal Settlements in the Tropics*, Advances in 21st Century Human Settlements,
https://doi.org/10.1007/978-981-13-7307-7_1

environment, transportation, taxation and budgetary policies has caused urban planning and planners to fail often to promote a more inclusive urban order, with discriminatory urban planning regulations based on unrealistic technical standards that do not take into account the socio-economic realities determining the conditions of access to land and housing.

1 A Different Approach to Tackle the Same Problem

The view and response of cities and governments toward the question of informal settlements has evolved over time, from the eviction, slum clearance, relocation and resettlement practices usually implemented before and during the 60s to the supportive practices of sites and services and participatory on-site slum upgrading implemented between the 70s and 90s, and in the last two decades the generation of citywide upgrading programs that define the future urban configuration of informal settlements from a resilient citywide perspective (UN-Habitat 2012).

Although the first approaches are still used in many countries, and informal settlements and settlers are still stigmatised in many cases, a new approach began to be clearly, and officially, adopted in September of 2000, when all the member countries of the United Nations signed the Millennium Development Goals (MDGs) declaration. The document proposed eight goals and committed nations to a new global vision with the aim of reducing extreme poverty by 2015. Specifically, in relation to informal settlements, Goal 7.D targeted "to have achieved a significant improvement in the lives of at least 100 million slum-dwellers by 2020" (United Nations 2015).

In this document, and in the 2010/2011 State of the World's Cities Report, the halfway document towards the accomplishment of the MDGs, UN-Habitat demonstrated a shift in the attention to informal settlements and a clear support for the on-site slum upgrading policies.

Building on the results of the MDGs, on the improvements achieved in the quality of life of millions of informal settlement dwellers, as well as on the biased approaches to informality by the Sustainable Development Goals (SDGs), the New Urban Agenda included in 2017 a clear statement towards the prioritisation of regenerating and upgrading on-site slums and informal settlements avoiding spatial and socio-economic segregation and gentrification, preventing and countering the stigmatisation of specific groups, and integrating informal settlements into the social, economic, cultural and political dimensions of cities (UN-Habitat 2017a). This statement was reflected in Goal 11, that established the objective of making cities and human settlements inclusive, safe, resilient and sustainable ensuring access for all to adequate, safe and affordable housing and basic services and upgrade slums by 2030 (United Nations 2017b).

In other words, in the past, planning, as an expression of the lack of political will, used to neglect and reject the informal city, pushing and forcing the less favour populations to settle at the urban peripheries and the edge of towns, disconnecting them from jobs, services and the basics infrastructure networks just by not providing decent access to secure land rights for housing. However, it can be argued that the new approaches to the regeneration or upgrading informal settlements take them not as the problem but as part of the solution. Regeneration or slum upgrading is understood, in this case, as the improvement of informal settlements and their integration into larger urban systems, far beyond the violent demolition and social displacements of previous approaches, as defined by Gouverneur in his chapter.

As explained in Chap. 2, the concept and practice of urban renewal that prevailed in the post-world war period in many European countries has been shifted into on-site slum upgrading and informal settlement regularisation strategies in the current developing world, aiming to resolve the challenges of the access to basic infrastructure such as water, sewerage and drainage and the formalisation of land tenure in favour of present residents and to ensure improvements in their resilience and quality of life.

2 The Phenomena of Informal Settlements in the Tropics. Confronting Asia and Latin America

Although no one solution fits all, and the upgrading approaches cannot be uniformly applicable to all the informal settlements, this book draws on the experiences of similar approaches to on-site slum upgrading and the regeneration of informal settlements in two different geographical and cultural contexts to contrast the impacts and to find similarities where to learn from.

More precisely, the book focuses its attention on the tropics geographical area that surrounds the earth's equator within the latitudes of the Tropic of Cancer in the north and the Tropic of Capricorn in the south. The interest in this area relies on the fact that the tropics is home to 40% of the world's population, and it is where the urbanised population rate has increased globally faster in recent decades, from 30.5% in 1980 to 45% in 2010, while in the rest of the world the urban population rate increased from 44.3 to 56.2% for the same period. The regions in the tropics with the most notable urban population increases were South-East Asia (24.1–47.2%), South America (65–81.4%), and the Caribbean (50.5–65.7%). South and South-East Asia together accounted for almost half of the urban population growth in the tropics since 1980. Also, it is important to note that South America, Central America and the Caribbean have been the most urbanised regions in the tropics since 1990, but urbanisation rates are expected to increase rapidly in South Asia and South-East Asia through 2050.

The result of this fast urbanisation in the tropics also has consequences for the population's living conditions. According to current data, more than two-thirds of the world's population living in extreme poverty are located in the tropics. The proportion of the urban population living in slum conditions is higher in the tropics: 46%, compared with 24% in the rest of the world. And in absolute terms, there are almost 470 million slum inhabitants in the tropics, compared to 460 million in the rest of the world (State of the Tropics Report 2014).

On the contrary, the most significant declines in urban population in the tropics living in informal settlements have taken place in Eastern Asia, South-Eastern Asia and Southern Asia, with an average of 13%. In Latin America and the Caribbean region, the proportion of the urban slum population declined by 9%, making them the regions with the lowest prevalence of informal settlements (United Nations 2015).

In other words, urbanisation is an extremely relevant process in the tropics, and more specifically in Latin America and Asia, but so is how informality seems to have been tackled efficiently in these two regions. This book aims to review some of the urban regeneration and on-site slum upgrading policies, approaches and strategies implemented in informal settlements in Latin America and Asia in order to find points of connection and similarities between the two cases, which can help in the formulation of adequate resilient planning to curb the expansion of informal settlements.

3 Some Reflexions About Regeneration Actions to Improve Informal Settlements

According to UN-Habitat, it is likely that by 2025, 1.6 billion people will require adequate and affordable housing. Improving the living conditions in informal settlements is indispensable to guarantee the full recognition of the urban poor as rightful citizens, to realise their potential and to enhance their prospects for future development gains. Therefore, on-site upgrading of informal settlements becomes an essential approach to move the world towards a rights-based society in which cities become more inclusive, safe, resilient, prosperous and sustainable (UN-Habitat 2016).

This book aims to be a contribution to the debates on on-site upgrading and regeneration strategies and approaches that can provide alternative systems to access housing for all urban residents. And one of the main messages and aims of the publication is to suggest underlining specific urban actions focused on the upgrading or regeneration of informal settlements as ways to improve the living conditions in these areas, while also building more equitable and resilient cities. The book structure takes inspiration from the urban action framework against inequality suggested by UN-Habitat/CAF (2014), understanding that inclusiveness and equity are indicators of improved and regenerated urban environments. The framework by United Nations explores possible relationships between some local development policies linked to the city structure under four components: spatial connection, social cohesion, capacity building and institutional coordination. The book takes these components as the general basis to organise in four parts the dissection of eight paradigmatic and pioneering case studies and initiatives that have developed and implemented these strategies. In each part of the book, two chapters explain selected visions in tropical Asia and Latin America under the same approach and strategy.

Although the book does not aim to be exhaustive in the cases and the proposed strategies, the selected references constitute some of the best practices in on-site upgrading. Therefore, the book provides an opportunity to show and evaluate their success as well as those aspects that might not be necessarily considered as entirely effective. In this sense, one of the main lessons learned from the book is that despite that we can come up with a systematic strategic approach to similar issues for different urban environments, it can be concluded that each action has to be adapted to the specific context and therefore, although it might seem obvious, there is no unique solution that fits all the disparities. Although informal settlements in Latin America and Asia share pretty much the same problems, have similar processes, and can be understood under the same strategic approaches, they cannot be tackled by generalised solutions that replicate the same policies. The notable differences in the stage of development that we have observed between the Latin America and Asia cases, the first showing a more consolidated status than the latest, coupled with different legal frameworks, cultural backgrounds, economic constraints, and political agendas, require particular sensitive and rigorous approaches in each case. It is precisely this rigour that the book takes from each of the presented cases to suggest four actions to improve the living quality in informal settlements.

In Part 1, the book contributes to the discussion of the action of urban regeneration through infrastructure and spatial connections. In many cases, a lack of adequate urban planning, weak regulatory frameworks and inefficient policies have resulted in informal urban patterns of occupation. This section provides a view on how the improvement of spatial connections can establish physical links for better accessibility, aiming to reduce the gaps and imbalances between informal settlements and the basic means for living.

In general, we can say that the urban regeneration approach through spatial connections aims to facilitate access to areas where employment opportunities, education, healthcare, culture and services are located, which will result in more integrated and inclusive physical urban environments. Thus, spatial connections and social cohesion are keys to creating equitable environments where all citizens have the same opportunities.

Chapter 2 elaborates on the street-led citywide slum upgrading approach as an innovative strategy to achieve the social and spatial inclusion of informal settlements and their residents into the city where they are located. The author argues that this approach constitutes an incremental development strategy to achieve the physical and spatial integration of slums into the nearby districts, and the city as a whole, by defining the network of streets, patterns of circulation and accessibility to and from the settlements. Practical examples in Brazil show the compelling results of this approach in achieving greater spatial connectivity and physical integration. Still, they also reveal the challenges when it comes to incorporating slum dwellers into the city's social, economic, institutional and city management environment.

In Chap. 3, the contribution to the discussion of on-site upgrading through spatial connections is made by evaluating the impacts of providing low-cost physical infrastructure in slum-upgrading projects. The case of the Kampung Improvement Programme (KIP) implemented in Indonesia, the world's first on-site urban slum upgrading project according to UN-Habitat, is dissected to study the physical, social, economic and environmental effects, benefits, successes, drawbacks, and failures. The practical example of the KIP shows a very significant impact on large segments of the population, suggesting fast, simple, replicable and low-cost regeneration tools to increase inclusiveness, as well as possible actions for a resilient urban regeneration of informal settlements.

Part 2 of the book elaborates on the provision of public space as a tool for urban regeneration. Several instances indicate that public space is a tool to create cohesion and encourage the development of social capital. Public spaces also aid in forging an identity and creating a sense of belonging, in addition to facilitating coexistence and the development of solidarities. Urban regeneration through the improvement and provision of public space—as an instrument of social cohesion—promotes a sense of citizenship and generates a sense of belonging to the community that reduces the possibility of conflict. It builds the capacity of civil society to organise its own support networks to address gaps while improving quality of life and environmental conditions, which results in better health for the population.

Chapter 4 contributes to the knowledge of on-site upgrading approaches by formulating a sustainable and replicable model for slum revitalisation based on the potential of public space as an urban regeneration strategy to establish long-term visions for the future of informal settlements. The chapter posits spatial analysis as a preliminary stage and a method in formulating structured revitalisation plans for informal settlements. Using Manila as a case study, the section explains how informal settlements work in order to establish alternative urban regeneration strategies that are socially, economically and environmentally sustainable.

Chapter 5 explains and highlights one of the most significant urban regeneration planning tools used in the city of Medellín to tackle informality: the Proyecto Urbano Integral (PUI) [Integral Urban Project]. Looking at the experience of the first PUI, tracing its origins, and considering the practices and experiments developed over three decades to address informal urban settlements, this chapter analyses the impact of the PUI as a planning tool and looks at how the use of public space in Medellín became the foundation for one of the main urban upgrading strategies. A decade after their implementation, the article contributes with the understanding of the strategic role of PUIs in establishing a method for urban regeneration in informal areas by providing structured public space, as well as the value of this planning tool as a pilot and replicable experience throughout the city.

Part 3 of the book brings to the table the discussion of urban regeneration through education and capacity building. It has been observed that education levels are inversely proportional to the level of informality in the labour market. This makes important to improve opportunities for access to decent and formal employment through training, involving the development of vocational training programmes based on demand from both labour and output, the implementation of transfer programmes conditioned to schooling, and the creation of facilities for comprehensive learning and social development. Learning from the Latin American experience, where informality is the main form of urbanisation, planning laws have been enacted or revised, revealing the importance of addressing informal urbanisation with proactive involvement from the public sector. This has given rise to the phenomenon where innovative planning, design and managerial paradigms have been brought into encourage community engagement.

Chapter 6 explains this approach with reference to two emblematic Latin American case studies—Caracas in Venezuela and Medellín in Colombia—which focus on the upgrading or rehabilitation of informal settlements, pointing out their achievements and limitations, revealing the importance of working on the ground, hand-in-hand with the communities, and the need for committed political support and efficient municipal management. The chapter contributes by providing innovative planning and design solutions for the urban renewal of informal settlements in the form of flexible design components, adaptable to different site conditions, which will provide informal cities with higher levels of performance and help residents attain decent living conditions.

Chapter 7 discusses a participatory planning proposition used in upgrading informal settlements by reviewing approaches to identify the failures of top-down processes and demonstrate the challenges of bottom-up processes. The chapter details a participatory planning approach that can achieve a balance between the bottom-up and top-down process approaches. The author's contribution suggests using flexible approaches to the upgrading of informal settlements, bringing together participation and management through a hybrid model of top-down and bottom-up. This hybrid model allows for design creativity, inclusive participation and feasible implementation of the project to improve the building capacity of the community.

Finally, Part 4 contributes to the discourses about integrated planning and governance, and urban regeneration through institutional coordination. Institutional coordination refers to the capacity to balance various initiatives intended to combat inequality, driven at different governmental levels and in different functional areas of a single level of government. The improvement of institutional coordination is encouraged through the development of synergies between the initiatives promoted locally and nationally. The idea is to build the necessary critical mass for policies to have a transformative effect, which ensures the continuity of the policy beyond government cycles. Political agendas and the prioritisation of resources generally come to light in an unequal way between different municipalities and regions. The leadership at every level of the government is a crucial factor in creating a critical mass of local politics, which can be more dynamic and oriented to demand.

Chapter 8 contributes to the discussions on governance and institutional coordination in the Latin American and the Caribbean (LAC) region and suggests governance reform solutions for improving socio-spatial justice in formal and informal settlements by analysing current conditions and learning from cases of different cities and countries in the region. The chapter indicates that dysfunctional governance and institutional coordination are two of the aspects hindering improvements in socio-spatial justice in the LAC region. This is a crucial issue in a region with one of the most substantial levels of informality and inequality in the world and where poverty rates are again expanding after showing encouraging retreats in the previous decade. The chapter sheds light to some of the governance challenges including, unclear and unstable institutional structures for decision-making, insufficient mechanisms for citizen participation, and lack of political will and appreciation for the benefits of planning.

The final chapter, Chap. 9, presents the case of Thailand and contributes to the discussion on how good community-driven planning and governance for the development of the low-income population can be a powerful tool to regenerate informal settlements from an inclusive perspective. The article argues for the role of the government in uplifting the low-income communities and assisting them in becoming self-sufficient. Studying the Baan Mankong Programme as one of the few initiatives worldwide that address the upgrading of informal settlements at the national level using a community-driven approach, the chapter suggests a possible framework based on adaptive governance inspired by people's initiatives and responding to their needs. A flexible and diverse framework to provide assistance, resources and infrastructure to the low-income population, building capacity and promoting self-sufficiency.

References

Abbott J, Douglas D (2001) A methodological approach to upgrading, in situ, of informal settlements in South Africa. Water Research Commission No. 786/2/01, Pretoria

Fernandes E (2011) Policy focus report: regularization of informal settlements in Latin America. Lincoln Institute of Land Policy

James Cook University (2014) State of the tropics report. Available at https://www.jcu.edu.au/state-of-the-tropics/publications/2014. Accessed on 13/02/2019

Njamwea MM (2003) Upgrading informal settlements by securing public spaces: case study of informal settlements in Blantyre City, Malawi. National Institute for Geo-Information Science and Earth Observation Enschede, The Netherlands

Payne G, Davidson F (1983) Urban projects manual: a guide to preparing upgrading and new development projects accessible to low-income groups. Liverpool University Press

UN Habitat CAF (2014) Construcción de ciudades más equitativas. Políticas públicas para la inclusion en América Latina

UN-Habitat (2003) Chapter 1: Development context and the millennium agenda. The challenge of slums: global report on human settlements (Revised and updated version, April 2010). Original version is available at www.unhabitat.org/grhs/2003

UN-Habitat and UNESCAP (2008a) Urbanization: the role the poor play in urban development. In: Housing the poor in Asian cities, vol 1

UN-Habitat and UNESCAP (2008b) Land: a crucial element in housing the urban poor. In: Housing the Poor in Asian Cities, vol 3

United Nations Human Settlements Programme (UN-HABITAT) (2012) State of the World´s Cities 2012/13. Prosperity of Cities, United Nations Human Settlements Programme

UN-Habitat (2014) Streets as tools for urban transformation in slums: a street-led approach to city-wide slum upgrading. Available at http://mirror.unhabitat.org/pmss/getElectronicVersion.aspx?nr=3552&alt=1

UN-Habitat (2016) World cities report 2016. Urbanization and development: emerging futures. UN-Habitat, Nairobi

United Nations (2015) The Millennium Development Goals Report

United Nations (2017a) Transforming our world: the 2030 agenda for sustainable development

United Nations (2017b) New urban agenda

Infrastructure. Urban Regeneration Through Spatial Connection

Improving spatial connection, establishes a link between land use and accessibility, eliminates the imbalances between residential areas and work places and reduces the gap between informal and consolidated areas. In general, urban regeneration through spatial connections facilitates the access to areas where employment opportunities, public facilities and services are located, thus limiting the territorial inequality.

Street-Led Citywide Slum Upgrading: Connecting the Informal and the Formal City Through Area-Based Planning and Infrastructure Improvement

Claudio Acioly

Abstract This chapter illustrates the street-led citywide slum upgrading approach as a strategy to achieve the social and spatial inclusion of slums and their residents into the city where they are located. The street-led approach draws on the area-based plan, the urban layout design and the street networks. This chapter argues that street-led citywide slum upgrading constitutes an incremental development strategy to the achieve physical and spatial integration of slums into the nearby districts and the city as whole. The practical examples of Rio de Janeiro and São Paulo show the compelling results of this approach in achieving greater spatial connectivity and physical integration, but they also reveal the challenges to incorporate slum dwellers into the city's social, economic, institutional and city management environment.

Keywords Informal settlements · Slums · Street-led slum upgrading · City-wide slum upgrading · Favelas · Sao paulo

1 Preamble

This chapter focuses on the planning and implementation of streets as the foundation for an area-based plan to trigger urban transformations in slums and informal settlements. The case studies referred to in this article substantiate the argument that the

[1]An area-based plan is an urban layout plan, also called a settlement layout design, that guides the future development of a slum, informal settlement or neighbourhood. It physically defines the spatial structure and urban configuration of the settlement. It also includes land-use patterns, land subdivisions and the street network, including the indications for the demolitions and relocation of buildings and housing units necessary to open new and/or improve existing streets. The area-based plan is the result of a physical planning intervention led by the street network, which is essential for the execution of the infrastructure networks, pipes, power lines, drainage, pavements and streets for the circulation of people, vehicles and public transport, which connects the settlement to the city where it is located.

C. Acioly (✉)
GIZ & EU—IUC-LAC, Brasilia, Brazil
e-mail: claudio.acioly@iuc-la.eu

O. Carracedo García-Villalba (ed.), *Resilient Urban Regeneration in Informal Settlements in the Tropics*, Advances in 21st Century Human Settlements,
https://doi.org/10.1007/978-981-13-7307-7_2

area-based plan[1] not only helps to define the future urban layout of these settlements but also integrates them physically and spatially into the rest of the city by defining the network of streets, and patterns of circulation and accessibility, to and from the settlements. This is called the street-led slum upgrading strategy, an approach to the urban renewal of informal settlements. The consolidation of streets and pathways in the territory is vital for the incremental improvement of infrastructure provision, fostering greater accessibility and laying down the basis for future land tenure regularization. Overall, residents acquire greater accessibility and connectivity to the city, and their places of residence are provided with access to infrastructure and basic urban services. It helps to lay down the foundations for the formalization and regularization of these settlements and their full insertion into the urban management systems that govern cities.

The concept and practice of urban renewal and urban revitalization strategies that prevailed in the post-World War II period in many European countries established the model for slum upgrading and informal settlement regularization strategies in the developing world. These strategies aim to resolve two paramount challenges: on the one hand, the access to basic infrastructure such as water, sewers and drainage; and, on the other hand, the formalization of land tenure in favour of present residents.

In order to achieve these objectives, it is fundamental to implement a settlement layout plan in order to reverse the process of spatial and residential exclusion affecting these areas and their residents, instead connecting the settlements to the rest of the city through the street and infrastructure networks.

A street-led approach to slum upgrading addresses these challenges in an incremental manner by promoting an urban restructuring strategy that uses streets as the primary physical vector through which water and sewer pipes, power lines, and drainage systems, and public life in cities are laid. Examples from cities in Brazil, Colombia, Venezuela, Argentina and Uruguay demonstrate the strength of this approach. Streets become a common asset and a vital element in enabling the spatial and physical integration of these settlements into the rest of the cities where they are located. In randomly and densely occupied slums, where accessibility and lack of trunk roads are common problems, the street-led approach becomes transformative because it triggers the opening of new streets or requires the improvement and consolidation of existing streets, which calls for the demolition of buildings and the resettlement of residents. The approach usually adopts a resettlement strategy that recognizes the rights of residents and relocates them to nearby areas or to locations within the settlements where they live. Thus, the implementation of a street-led approach to slum upgrading will inexorably be associated with resettlement and relocation, triggering strong citizen participation in the prioritization of the streets and the future urban layout configuration of their neighbourhood.

This article cites evidence to argue that the development of streets is essential to any slum upgrading intervention, and it draws on UN-Habitat's strategy paper that promotes the street-led approach to urban transformations.[2] A meticulous analysis of the existing urban configuration and land uses, and the identification of the existing

[2] Acioly (2014), UN-Habitat (2012a).

patterns of circulation used by residents in slums and informal settlements, becomes part and parcel of the urban planning process. This analysis enables strategic spatial planning decisions and supports choices to be made regarding the selection of streets for improvement and new ones that need to be created and/or widened to facilitate accessibility and the most efficient use of land and infrastructure. This is likely to propel economic activities with positive impacts on the settlement and in the city as a whole—not mentioning the significant impact on the lives of residents. Land allocation for resettlement and relocation is also another critical element in spatial planning decisions when implementing a street-led slum upgrading strategy.

2 Slums: Facts and Figures

In its seminal global report[3] launched in 2003 after the adoption of the Millennium Development Goals-MDGs and the agenda to improve the lives of slum dwellers, UN-Habitat estimated that nearly 1 billion people were living in slums. The majority were in developing regions, which represented 43% of the urban population. This was the first global assessment of slums, and if no concerted actions are undertaken by governments and the international community, projections indicate that by 2033 the population living in slums will increase to 2 billion people. In order to measure progress and monitor the MDG 11/7 on improving the lives of 100 million slum dwellers, the report adopted an international operational definition of slums combining physical and legal dimensions, such as: inadequate access to safe water; inadequate access to sanitation and other infrastructure; poor structural quality of housing; overcrowding; and insecure residential status. A monitoring system was put into place to enable the annual report for MDGs.

In 2010, data from the State of the World Cities Report[4] suggested that the proportion of the world's urban population living in slums decreased from 46.1% in 1990 to 32.7% in 2010, while in absolute terms the population was increasing. In 2013, UN-Habitat issued its urban statistics during the 24th session of its Governing Council, published in its global report,[5] which showed that one out of four people living in urban areas in the world were living in slums. In Africa, one out of every two people living in urban areas were living under these conditions. More recently, the estimates provided by UN-Habitat[6] show that the proportion of the urban population living in slums has further decreased to 29.7% in 2014. However, in absolute numbers the total of slum dwellers is still increasing, and in 2014 over 880 million people lived in slums, compared to 791 million in 2000 and 689 million in 1990.

[3]UN-Habitat (2003).

[4]UN-Habitat (2010d).

[5]UN-Habitat (2012b).

[6]UN-Habitat (2016).

3 Revisiting the Foundations of the Street-Led Slum Upgrading Strategy: The Area-Based Plan

Slums and informal settlements are mostly disconnected from the urban management and planning systems that govern cities. Usually, their residents remain spatially and socially segregated by and large. When compared to the rest of the city, these settlements often remain as unplanned enclaves that are poorly serviced by basic infrastructure, lacking streets and public spaces, and condemned to the adversities derived from the lack of connectivity with the city's urban fabric.

During the last 50 years, slum upgrading interventions have addressed these problems in various ways. We identify a historical sequence of interventions moving from improvement of infrastructure, towards improvement of infrastructure and land regularization, and finally towards improvement and full urbanistic and spatial integration.

Also, global research[7] evaluating policies targeting informal settlements and slums, which sponsored the Urban Management Programme during the 1990s, also revealed four major approaches that illustrate the foundations and evolution in slum upgrading interventions: the first, the legalization of land tenure focusing on land-use and property regularization; the second, settlement upgrading meant to achieve a rational urban layout pattern and optimal land use, which includes re-blocking and densification, with eventual relocation of residents; the third, with an overly ambitious settlement upgrading, aimed at infrastructure provision to overcome years of neglect and poor provision of services; and, finally, a preventive approach that encompasses an incremental provision of primary infrastructures combined with the recognition and incorporation of these settlements into formal systems, accepting a gradual compliance with norms and laws and the progressive formalization of properties and urban restructuring. The area-based plan and the design of the urban layout, with its network of streets, were intrinsically associated with these approaches.

Following these approaches, we also identify four generations of street-led upgrading projects. The first generation of projects focused on infrastructure improvement and sanitary conditions of the settlements, without much concern for the urban structure of the area. A second generation included elements of community participation and participatory planning, coupled with efforts to ensure security of tenure for the inhabitants and their involvement in infrastructure improvement processes with little concern for the urban configuration of the settlement. A third generation of projects, emerging particularly from the 90s onwards, focused on integrated settlement planning and development, led by a vision for the city as a whole that focused on integrating slums into the city, moving away from a project basis towards citywide program interventions and paying greater attention to the settlement layout, the urban plan, the land subdivision and the outline of the public and private domains.

[7]Durand-Lasserve and Clerc (1996, pp. 233–273).

This last generation of projects adopted an approach to improve the provision of basic infrastructures such as water, sewers, electricity or drainage, implementing them together with improvements in the physical accessibility and spatial connectivity of the settlements in relation to nearby neighbourhoods and the rest of the city. This was guided by an area-based plan that defined the future urban configuration of the settlements. Finally, a fourth generation of street-led projects emerged at the beginning of the twenty-first century combining these approaches with a strong focus on the urbanistic and land property regularization, sustained by a rights-based approach that recognizes the rights of residents over the land they occupy. Very often, as in the case of Brazil and Colombia, this was associated with the enactment of national legislation regulating land rights in urban areas. In this case, the aim was to lay down the foundations for the gradual transformation of slums into formally consolidated neighbourhoods of the city. It goes without saying that the recognition and commitment of member states of the United Nations in 1996 to the full and progressive realization of the right to adequate housing, with the adoption of the Habitat Agenda by governments and heads of states attending the UN Conference Habitat II in Istanbul, had a strong influence and provoked shifts in policy and approaches towards slums and informal settlements.

It does need to be said that the adoption of street-led strategies promoting the opening of new streets and pathways for the improvement and consolidation of existing streets is not new. This approach has long been part of the common practices of slum upgrading and the renewal of informal settlements aimed at the creation of public pathways for the execution of infrastructure networks and for the improvement of accessibility and the overall quality of life and living conditions in slums. The new element that we recognize is the formulation and implementation of the area-based plan, drawing on a strong urban planning and design approach which creates a network of streets and public spaces that reconnects slums to the urban fabric of the city. The area-based plan defines the public and private domains in the settlements and ultimately the new urban configuration and street networks. This is the cornerstone for local development processes, infrastructure improvements, land tenure regularization, and future property and housing improvements fuelled by the residents' resources.

4 Defining the Street-Led Slum Upgrading Approach and Its Goals

The street-led approach sees streets not only as a vehicular road but as a route for incremental urban transformation which aims to integrate slums spatially, physically, socially, juridically and economically into the overall city development strategy. This approach highlights the social and economic dimension of streets as part of the public domain.

The street-led approach to slum upgrading which was adopted by UN-Habitat in 2012[8] brings all these elements into a single strategy. It transforms the renewal of informal settlements and slums into an urban restructuring strategy that makes use of streets not only as the pathway for mobility and accessibility through which pipes, power lines, and drainage systems are laid out. It posits the streets as a common asset and as part of the public domain, where social, cultural, economic, and legal dimensions come into play.[9] Streets thus become vital elements of the area-based plan to transform slums and to improve the quality of life and ultimately connect these areas with the rest of the city. This is highly needed in densely and haphazardly occupied informal settlements, where the lack of streets and public spaces place an additional hardship on people's lives.

The adoption of a street-led approach, which includes the area-based urban plan, requires taking a critical look at the demolition of buildings and the selective identification of families/buildings that must be relocated in order to create the future urban layout of the settlement. This makes space to lay down the infrastructure networks and improve accessibility via the openings or consolidation of streets and pathways for pedestrians and vehicles where needed and where possible. Given the irregular and unplanned character of the urban layout of slums, the provision of basic infrastructures without resolving the urban spatial structure and street pattern of the settlements has proven to be ineffective and costly and should be discouraged. In this sense, the area-based plan is critical to defining the public domain through which the networks of infrastructure will be implemented, and it is a way to establish a strong connectivity with the rest of the urban fabric of the city, generating greater accessibility and opportunities for further property regularization (Acioly 1992; 1993).

Several examples of projects in the favelas during the Favela Bairro Programme in Rio de Janeiro demonstrate this.[10] Opening streets, or reinforcing and improving existing streets and accesses, is a sine qua non in the renewal and upgrading interventions of slums and informal settlements,[11] with the objective of creating an urban configuration that physically and spatially integrates slums into the overall city planning and management, and regains spatial connectivity while fostering the urban economy and the physical regeneration of the area.[12]

Selective demolitions and relocations derived from street-led slum upgrading will have a direct impact on levels of urban density and patterns of land occupation and regularization.

---

[8]UN-Habitat (2012a).

[9]Gehl and Gemzoe (2006).

[10]Conde and Magalhães (2004).

[11]For a more detailed analysis of public spaces, streets, and public life in cities, see Favro and Inger (1994), Ghel et al. (2006), Gehl (1996).

[12]UN-Habitat (2012a).

5 The Practice of Street-Led Slum Upgrading

The foundation of the practice of street-led slum upgrading lies at the core of urban spatial planning at the settlement level. At first, a careful analysis of existing patterns of streets and pathways, circulation, social uses of space and the existing urban configuration should lead to strategic design decisions and choices defining approach in phases. It helps to identify existing patterns of streets and pathways, some of which can be upgraded and consolidated. The streets initially selected for improvement will be those that are likely to bring the best outcome in terms of development opportunities, poverty reduction, accessibility, efficiency and optimization of land use, and generation of urban and property value. The choice of a particular street, followed by others incrementally, will determine the logic of land subdivision and future property regularization, and maximize the economy of those streets as a function of slum upgrading, which will likely propel development and income generation activities with impacts on the city as a whole.

The street-led strategy to citywide slum upgrading is consistent with property rights and the right to adequate housing, which enshrines the right not to be forcibly evicted. While it recognizes the need for demolition and the relocation of some residents in order to provide room for a better and more optimal urban configuration of the settlement, it is also unwaveringly grounded in the rights of individuals and households through participatory enumerations and solid community involvement. This strategy addresses the need to provide accessible and well-designed public spaces, which are important elements in the agenda to improve quality of life and social interaction within the public domain. It is worth noting that street-led slum upgrading is closely associated with the provision of public space, which contributes directly to the achievement of the Sustainable Development Goals (SDG) Target 11.7, aimed at the provision of universal access to safe, inclusive, and accessible green and public spaces, in particular for women, children, senior citizens and persons with disabilities. In this respect, it also contributes to the SDG 5, calling for the achievement of gender equality and empowerment of all women, which is also linked to SDG 11 through access to and safety in public spaces, access to and use of basic infrastructures, and participation in local governance and decision-making.

International examples show that it is feasible to implement this strategy, and the demolitions and relocations required for this purpose can be undertaken through processes of community mapping, enumeration, and participatory planning. In the cases of Ghousia Colony (Karachi), Tirana (Albania), and Bissau (Guinea-Bissau), once the advantages of street widening were discussed, people got involved in field surveys and in the demarcation of street boundaries and private plot boundaries. They voluntarily gave up portions of their plots and demolished and reconstructed the affected properties (Acioly et al. 2003). Relocations were also carried out in Karachi in a cooperative manner.

The case of the favela-Bairro Programme in Rio de Janeiro also involved the demolition of buildings and the relocation of residents to open up streets using criteria defined jointly by the Municipality and the Inter-American Development

Bank. The programme defined acceptable amounts of demolitions and relocations and the threshold for financing housing reconstruction so that upgrading would not be transformed into a resettlement project. The strategy was to keep relocation to the minimum necessary to realize the urban restructuring of the settlement and relocate residents to resettlement areas situated as close to the original settlement as possible, in order to minimize the adverse impact on residents' social and safety networks.

In Lusaka, Zambia, in one of the first large-scale slum upgrading and sites and services projects financed by the World Bank during the 1970s, road planning groups were organized to draw up demolition and relocation plans to be presented to the community, which included alternate alignments of roads for settlement upgrading, clearly showing the demolitions required in order to implement the agreed-upon road network. Once the plan was agreed, relocations were accommodated. Finally, in Medellin, Colombia, a negotiated process of demolition and selective relocation in the first decade of the twenty first century provided room for urban regeneration and for clearing rights of way and building-free sections of land for implementing the Metrocable system – a cable car public transport system to provide access to poor and low-income settlements located on slopes. Those affected were supported to relocate to new housing within the area or outside, by choice. Compensation was based on the amount required to buy a house, but lower than the cost of new social housing.

6 The Rationale and Benefits of a Street-Led Approach to Renewing and Upgrading Informal Settlements

A study carried out by UN-Habitat provides evidence showing a strong co-relation between high street connectivity index and high city prosperity index.[13] In other words, a well-planned and well-connected urban spatial structure promotes prosperity for all. Conversely, cities that have large parts of their urban structure as unplanned, subject to an informal and random-type of land occupation, are penalized on its city prosperity index. Thus, unplanned and informally occupied land that lacks streets and connectivity to the rest of the city—pockets of randomly occupied land—adversely affects the lives of all city residents. In another study, UN-Habitat provides evidence drawn from an international comparative analysis showing the amount of land allocated to streets and the number of street intersections per km^2. This study reinforces the argument that the amount of land taken for opening streets and public spaces is an important feature of the spatial plan and the ultimate urban spatial structure of cities. Cities that have adequate land that is used for streets and public spaces, and reveal greater connectivity, are cities that are more liveable and productive.[14] Scaling up street connectivity through urban spatial restructuring by adopting a street-led upgrading strategy is likely to bring positive transformations of

[13] UN-Habitat (2013a).

[14] UN-Habitat (2013b).

slums and informal settlements which are beneficial for informal settlers and slum residents, as well as residents of other neighbourhoods.

A street-led approach to slum upgrading also promotes an incremental spatial inclusion, residential inclusion and physical integration, which are likely to stimulate economic activity. Because street-led approaches strengthen the development and consolidation of streets, this also ends up attracting commercial activities, the opening and clustering of shops and services along these main access routes, encouraging private investment in housing and business improvements and triggering cultural identity and a sense of belonging amongst the residents. The adoption of street-led approach frees up land for and paves the way for universal access to basic infrastructures and helps to promote a sense of security, triggering planned urban development that supports the achievement of the Sustainable Development Goal 11.[15]

In addition to providing better accessibility and connectivity, and the optimal use of the networks of basic infrastructures, the street-led approach increases the sense of belonging because it facilitates the definition of the street addresses, names and locations of the inhabitants' places of residence. This allows residents to gain and claim a physical address and a postal code for the place where they live. The benefits of this approach are clearly demonstrated in the case of the favelas in Rio de Janeiro, where street addressing was a citizenship demand of the favela movement and a part of the city's urban development strategy.[16] Ultimately, this approach supports the realization of the right to the city by slum residents.

7 Reflections on Citywide Slum Upgrading from Latin America

Slum upgrading has a long track-record in Latin America. As stated by UN-Habitat[17] in its report on the street-led citywide slum upgrading approach, the experiences in Latin America provide lessons that have global significance. The report highlights the importance of building citizenship through integrated slum upgrading interventions, intertwined with participatory planning that addresses the physical, social, and economic problems found in the settlements and neighbourhoods linked to them.[18]

The Inter-American Development Bank (IDB) has been a key actor in the Latin American experience, sponsoring more than 20 slum upgrading programs in different cities in the region, leading to improvements in the self-built housing stock, better health indicators, and reduced vulnerability. As a result of these programs, housing

[15]The Agenda 2030 adopted in 2015 by the United Nations as one of the outcomes of the Rio+ Conference on Environment and Development laid down 17 Sustainable Development Goals (SDGs). The SDG 11 aims at making cities and human settlements inclusive, safe, resilient, and sustainable.

[16]Acioly (2001).

[17]UN-Habitat (2012a).

[18]Rojas (2010).

conditions and accessibility to workplaces were improved through investment in streets and the physical integration of slums. Improvements in potable water provision, access to sewage systems, and garbage collection also resulted in significant impacts when executed in conjunction with street works.[19]

In another study of 11 slum upgrading experiences completed in the mid-1980s in multiple countries in Latin America, the authors[20] show how investments in urban services delivered through streets generates private investment in home improvement carried out by residents using their own resources. Investments in infrastructure generated de facto tenure security and significant increase in property value. In addition, the research conducted by Durand-Lasserve, Alain and Clerc[21] shows us some case studies from various slum upgrading experiences during the 1990s, which demonstrated the suitability of an incremental development approach to urbanization and infrastructure provision as the foundation for an effective slum upgrading. Street paving, connected to minimum standards of infrastructure and service provisions, proved to be successful, which is consistent with the street-led approach to citywide slum upgrading that advocates for phased development, executed incrementally and focusing strategically on a few streets at first to deliver higher impacts and trigger subsequent development. By underscoring shortcomings of approaches aiming at high standards of infrastructure and legalization of land tenure via individual titles, the research conversely supports the incremental street- led propositions.

8 The Brazilian Experience in Street-Led Slum Upgrading: Favela Jacarezinho

Brazil has a long experience in the upgrading of informal settlements and slums.[22] The largest cities in Brazil, like Rio de Janeiro, São Paulo, Belo Horizonte, Recife and Brasilia have developed citywide slum upgrading programs, including the street-led approach associated with area-based plans as one of the strategies. These programs recognize the need for these strategies in the implementation of infrastructure networks, cables, pipes, drainage gutters and power lines, while showing that street-led citywide slum upgrading helps to improve the connectivity and integration of the settlements into the physical, economic, social and environmental domains of the city. When the upgrading is embedded in the urban planning process of the city, it becomes part and parcel of the detailed planning at the neighbourhood level. In addition, the proposal of an area-based plan fully connects the informal settlements with the planning of the surrounding neighbourhoods, providing a vision for future transformations and the foundations for the formalization of these areas, especially when it is executed as part of the participatory area-based plan.

[19]Brakarz et al. (2002, 67).

[20]Skinner and Wegelin (1987).

[21]Durand-Lasserve and Clerc (1996, pp. 233–273).

[22]UN-Habitat (2017).

The Favela-Bairro Citywide Slum Upgrading Programme was a ground- breaking and pioneer municipal programme implemented on a scale never seen before, promoting multiple upgrading projects in prioritized favelas under a well-established programme coordination system, which was part of an innovative institutional management framework[23] established by the Municipality of Rio and fully operational during the period 1992–1995. The slum upgrading projects carried out under the Favela-Bairro Program,[24] executed by the Municipality of Rio de Janeiro for more than a decade from 1993 onwards, adopted the street-led slum upgrading and the area-based plan strategies as the foundation for public intervention in the favelas, putting into practice innovative tools aimed at the conversion of favelas into formal city neighbourhoods. The urban layout plans, followed by street naming, allowed for strong citizen participation in formalizing the addresses and giving names to streets in the settlements and, by default, including them in the map of the city and the public registry of addresses. These were particularly important for creating a unique identity and an official postal address, which facilitated people's access to an official place of residence—one of the pre-conditions for acquiring credit in shops and commercial transactions. Not mentioning the inclusion in the city map. Public spaces in the form of small plazas, leisure spaces and sports facilities were implemented, closely linked with the opening and urbanization of streets, resulting in greater accessibility and an improved quality of life. Opening streets also allowed for policing and an extension of city services into the favelas, which was followed by the UPP[25] program and social development programs in more than 200 favelas.

The case of the upgrading plan of the Favela Jacarezinho is a good illustration the street-led approach. Jacarezinho is one of Rio de Janeiro's largest slums, with nearly 50,000 inhabitants and originally considered as a large favela, subject to a specific Favela Bairro Programme targeting large favelas (Figs. 1 and 2).

Following the model of public tendering adopted by the Favela Bairro Program for the individual projects in the favelas, the private architecture and planning firm PRODEC/ArquiTraço was commissioned to undertake the project design for Favela Jacarezinho. The goal was to lay out the settlement plan and an urban regeneration strategy that would include streets and public spaces, greater accessibility and connectivity with surrounding areas, lower building densities when possible, amongst other things. After the first plan was presented, the municipality requested the formulation of another variant of the plan that would produce more streets and squares, and greater accessibility in order to decrease the building density. The involvement of the residents and public participation was not always easy due to the actions of the drug barons in the area. The issues about how much relocation would be acceptable for laying out the proposed street network and how many houses were to be demolished to break up the very dense occupation for the creation of public spaces were the

[23]Acioly (2001b).

[24]Magalhães (2016).

[25]The Pacifying Police Unit (Unidade de Polícia Pacificadora also known as UPP) was pioneered in Rio de Janeiro with the goal of reclaiming territorial control in the favelas, which were controlled by drug dealers.

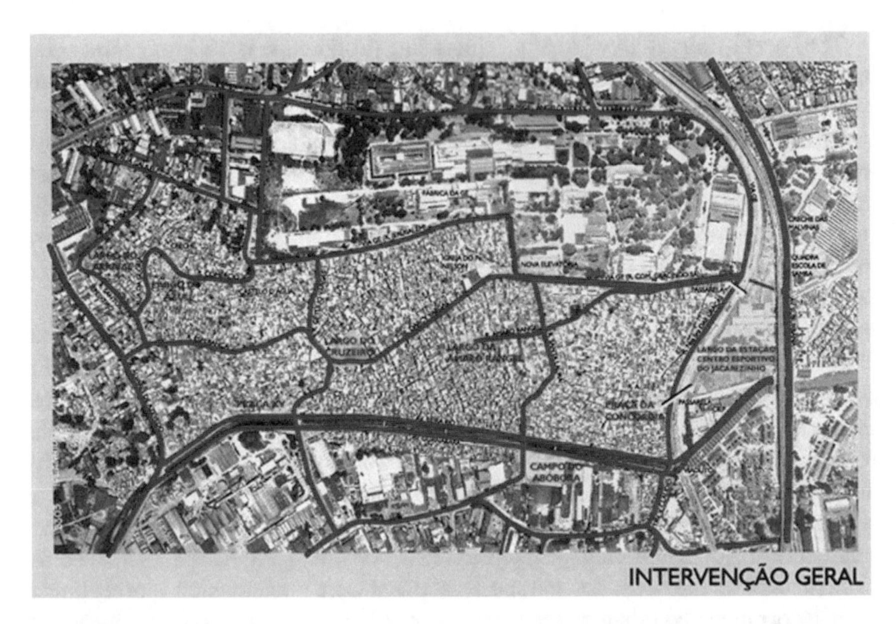

Fig. 1 Slum upgrading plan of the Favela of Jacarezinho *Source* Map elaborated by ArquiTraço and PRODEC. Courtesy of Solange Carvalho, team leader of the project

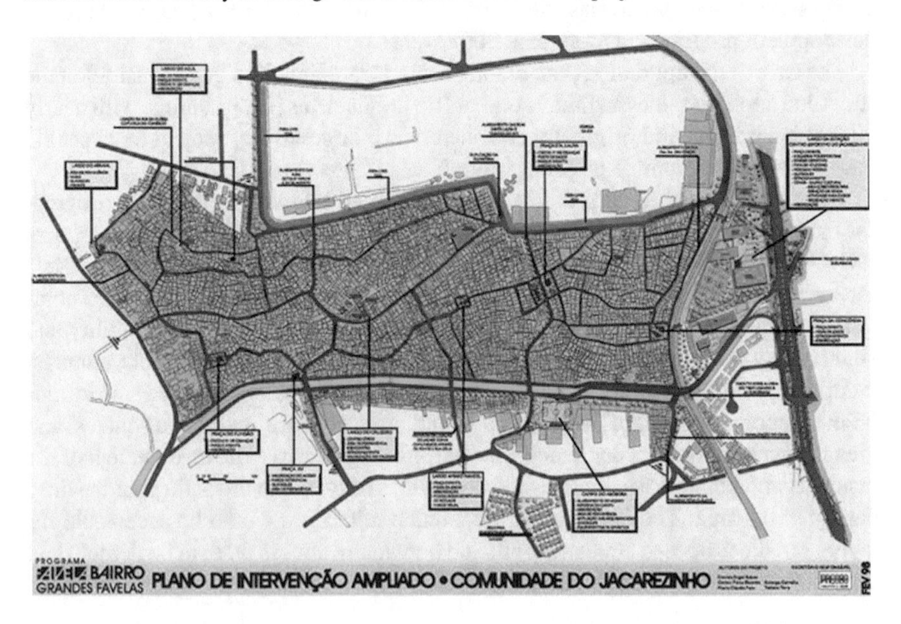

Fig. 2 Expanded slum upgrading plan of Favela Jacarezinho *Source* Map elaborated by ArquiTraço and PRODEC. Courtesy of Solange Carvalho, team leader of the project

subject of heavy debate. A partnership with the neighbouring GE Industry enabled the opening of an important access avenue linking two neighbourhoods through the settlement, which became one of the first land readjustment projects in the city.

Roads were designed to provide access and create public spaces where none existed before. These streets and spaces perform symbolic and physical functions of connections within the settlement and with the city, and they also allow the physical presence of the state—as police are able to physically move through the area for the first time, trying to control organized crime and drug trafficking, endemic in most favelas. The new roads provide an artery for the distribution of basic services, and they allow the creation of points of concentration for gathering and for the location of public amenities, many of which are shared with adjoining neighbourhoods, leading to greater inclusion. The naming and numbering of streets has brought the extension of city mapping into favelas for the first time and gives these dwellings a location and identity in the city, giving them a reality they did not have before. As a consequence of these actions, the housing values have increased in upgraded favelas. The programme is an example of slum upgrading without full land tenure legalization and for its use of the concession of right to use land in order to allow this program to take place. There was greater emphasis on the improvement of infrastructure and living conditions through opening roads and pathways, which has had the effect of increasing the security of tenure of favela residents.

Opening streets was the most difficult task to negotiate in the upgrading because of the steep terrain in most favelas, demolition of some houses, and financial limitations. However, an evaluation of the first two phases of Favela-Bairro by IDB shows that the multiple benefits far surpassed the difficulties.

9 Street-Led Citywide Slum Upgrading Programme in São Paulo

São Paulo is the largest city in Brazil, the largest city proper in the Americas and the southern hemisphere, and one of the top 10 largest metropolitan regions in the world, with a robust economy that contributes a significant share of Brazil's GDP. In 2008, the municipality of São Paulo had nearly 11 million inhabitants, the largest agglomeration in the São Paulo Metropolitan Region, comprised of 39 municipalities, where more than 20 million people live and work. Today, the city alone is estimated to house 12 million inhabitants.[26]

Nearly 10 years after the ground-breaking Favela Bairro Programme in Rio de Janeiro,[27] the municipality of São Paulo kicked off the development of a comprehensive urban development strategy in 2001, focusing on the poorest segments of

[26]http://worldpopulationreview.com/world-cities/sao-paulo-population/.

[27]The citywide slum upgrading-Favela Bairro Programme was launched in 1993 by the Municipality of Rio de Janeiro as an integral part of a broad municipal housing policy for the city, led first by an interim Municipal Housing Department.

the population.[28] In 2004, it embarked on the formulation of a sustainable municipal housing policy, closely associated with the establishment of a comprehensive information system that included geo-referenced mapping and cadastral information[29] about all the favelas and informal settlements in the municipality's territory, with detailed information about each settlement, households, housing enumeration, occupants' social economic profile, site risk assessment, etc. The municipal information system unveiled the existence of nearly 1600 favelas, more than 1000 irregular settlements and illegal land subdivisions, and nearly 1700 tenement housing areas (cortiços).[30] The favelas and irregular settlements, predominantly located in the peripheral areas, provided housing for 30% of the city's population,[31] the equivalent of about 3 million inhabitants. This study was the result of a collaboration with the Cities Alliance and helped the municipality to design well-informed and evidence-based policies and programmes.

Just like Rio de Janeiro did in the beginning of the 90 s, São Paulo also adopted a comprehensive municipal housing policy in 2007,[32] which included multiple programmes addressing the problems of slums and informal settlements, inner- city tenement housing, and irregular land subdivisions, amongst other issues such as tenure regularization and upgradation of public housing estates. The citywide slum upgrading programme[33] (2005–2008) led by SEHAB, the Municipal Housing Secretariat, was one of the largest of its kind in Brazil after Rio de Janeiro's Favela Bairro. The municipal housing policy remained in effect during the period 2005–2012 and provided important lessons for the practice of slum upgrading, as outlined thereafter.

The citywide slum programme drew on the lessons and experiences of the Municipal Program for the Urban Regeneration and Environmental Sanitation of the Catchment Area and Water Basin of the Guarapiranga lake/dam. This was a complex flagship program funded by the World Bank, which addressed a conurbation of informal settlements located within the jurisdiction of seven municipalities. Most of them originated from illegal land subdivisions, encroached onto the city's most vulnerable and sensitive environmental areas which provided the city with part of its potable water provision. The scale of the occupation and the level of the building consolidation made it nearly impossible to think of any other solution rather than the upgrading and urban regeneration of the area to prevent pollution and preserve the environmental resources, particularly the water catchment areas vital for the city. This was regarded as the first large-scale slum upgrading programme in the city of São Paulo.[34]

---

[28]Cities Alliance (undated).

[29]At first, the database was stored at www.habisp.inf.br and served for the municipality to formulate evidence-based policies and take well-informed policy decisions. Today, the information is accessible at: http://www.habitasampa.inf.br/.

[30]Prefeitura de São Paulo (2008a).

[31]Prefeitura (2008a), Cities Alliance (2008), Ibid.

[32]Prefeitura de São Paulo (2008b).

[33]Prefeitura de São Paulo (2008a). Cities Alliance (2008).

[34]França (2013).

One of the main lessons learned from the Guarapiranga Programme relates to the strategic importance of urban planning as an enabler of the integration and physical/spatial connectivity of the settlements with the surrounding districts, combined with the improvement of public space and the provision of public assets: streets, accessibility, and basic infrastructure services such as water, sewers, drainage, and street paving aimed at improving the quality of life and public services. Although it is not named as such, the street-led upgrading approach established the foundations for the land tenure regularization and registration and, by default, for the future consolidation of the settlements, while encouraging individual private investment in housing construction and home improvement.

The São Paulo citywide slum upgrading programme focused initially on 26 favelas, addressing 130,000 families—the equivalent of nearly 500,000 people. A planning and implementation methodology was developed to guide public interventions, and an area-based plan was formulated for each favela through a technical but also participatory process, involving the inhabitants via focus group meetings and general popular assemblies. The plans comprised a street network to facilitate accessibility and circulation of motorized vehicles and people on foot. There was a deliberate intent to enhance accessibility so that ambulances, garbage collection, police, postal services, as well as potable water, sewage and drainage network systems could be implemented as efficiently as possible. The explicit goal of the programme was to integrate the favelas into the city and provide quality public space, while improving inhabitants' accessibility to work places, schools, and health facilities to reverse the social and spatial segregation. The assumption was that the government would focus on the public assets, while housing and home improvement would be left entirely to the residents using their own means, savings and private resources. Spatial inclusion was meant to generate trickle-down effects in terms of social inclusion, accessibility, and better quality of life.

The Citywide Slum Upgrading Programme was part and parcel of the strategic master plan enacted in 2002 and provided the opportunity for the municipal government to adopt the instruments of the Statute of the City (Law 10.257/2001) regarding tenure regularization. The programme established a laudable institutional collaboration framework, aligning with the MDGs (water and slum improvement) and involving multiple institutions from the federal and state governments and international organizations such as the CDHU-Housing and Urban Development Company of the State of São Paulo, the Cities Alliance, the State Housing Coorporation-Cohab, and the PAC-Growth Acceleration Programme.

The lessons learned from the first seven upgrading projects reassured the assumptions of the programme and provided important feedback to support adaptations and refinements to the upgrading strategy. The settlement plan and provision of good quality public spaces strengthened the connectivity and integration within the nearby districts and neighbourhoods, giving indications of how to restructure the physical space and layout of the favelas. The plan was key in the physical/spatial integration

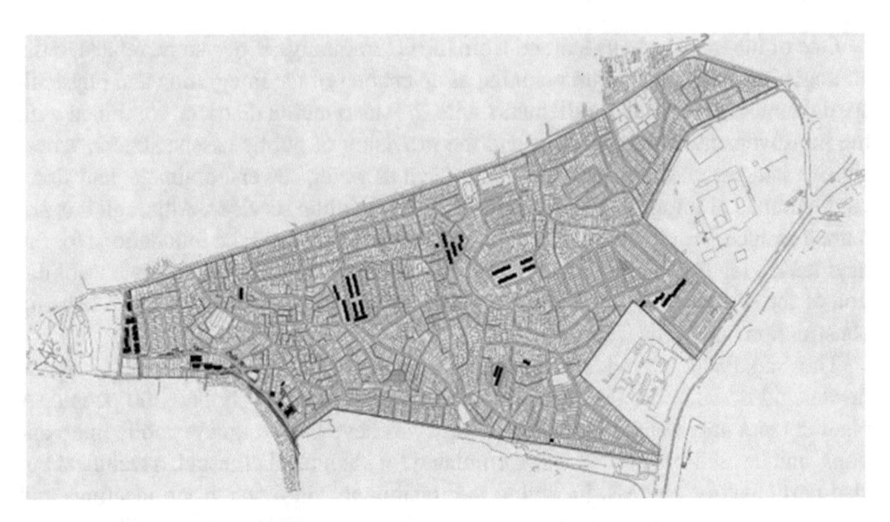

Fig. 3 Plan of Heliopolis *Source* Municipality of São Paulo

of the settlements into the city, which took into account the building stock produced by the inhabitants.[35]

The favela upgrading projects implemented under São Paulo's citywide slum upgrading programme demonstrated the adoption of a street-led upgrading approach which was built into the area-based plan, an urban layout design, and the street network. The urban layout design set the parameters for the plan and the street-led development process. This was the foundation for directing public investments into the public assets along the main streets, which the residents were neither able to access nor realize on their own. In other words, it focused on improving and/or (re)defining the street networks through which the drainage, sewage and water supply systems, as well as the electricity power lines and street pavement, were laid down. The (re)definition of streets required the demolition and later resettlement of inhabitants whose houses were located along the outlines of the street networks and pathways for infrastructures, as well as those located in risk-prone areas. The upgrading plans for Heliopolis, one of São Paulo's largest favelas, and Paraisópolis provide an illustration of the street-led citywide slum upgrading approach, where the street network and new housing are visible (Figs. 3, 4 and 5).

The ultimate goal of these projects was to integrate these areas into the spatial structure of the city and reverse the extreme level of social and spatial exclusion to which the residents of these areas were subject. It is worth noting that this approach is fully aligned with the Agenda 2030, and particularly with the Sustainable Development Goal 11, aiming at leaving no one and no place behind by supplying affordable housing for all and the upgrading of slums.

There are important lessons to be learned from São Paulo's experience with slum upgrading. First of all, slum upgrading is a key element of the municipal housing

[35]Prefeitura (2008a), Cities Alliance (2008), Ibid.

Fig. 4 Section K of Heliopolis Plan *Source* Municipality of São Paulo

Fig. 5 Favela Paraisópolis in Sao Paulo *Source* Maria Teresa Diniz. Manager of the Paraisópolis Upgrading Program. Municipality of Sao Paulo

policy to promote improvements in the overall housing stock accessible to the low-income population and in the access to public services. The provision of accessible, good quality and well-designed public spaces reinforces the physical and spatial integration of favelas into the city, enabling their residents to become citizens who fully enjoy the benefits of urban development and gain access to jobs, services, mobility, and so on. This is a result of the urban design of the settlement, which took into account social practices, as well as geological, physical, and morphological aspects of each favela. The design played an important role in defining the urban layout—the street networks combined with public spaces—which triggered the incremental transformation of favelas into an articulated and accessible neighbourhood.[36] Slum upgrading was thus a tool for urban transformation and for social and spatial inclusion. In that regard, the street-led slum upgrading became a tool to generate inclusive cities and neighbourhoods.

10 Concluding Remarks

Urban transformation in slums, if successful, should reveal an incremental development process through which informality and randomly occupied lands are gradually replaced by planned and deliberate actions intended to integrate slums into the formal governance systems that regulate urban land development in cities. The area-based plan and the accompanying street network lay down the foundations for this transformation, which includes social and spatial inclusion, land and property regularization, accessibility to basic infrastructure and an overall improvement in the quality of life of the inhabitants. This street-led transformation reveals the transformative potential of the approach to producing the inclusive city. In this respect, it brings about the following changes:

- the slum dweller becomes a citizen;
- the shack becomes a house; and
- the slum settlement becomes a neighbourhood.[37]

This transformative and time-bound path was highlighted by six slum upgrading experiences from different megacities in Asia, Africa and Latin America, which were presented during a dialogue sponsored by the Municipality of São Paulo and the Cities Alliance.[38]

The Brazilian experience, and in particular the experiences in Rio de Janeiro and São Paulo, provide important lessons for countries and cities that wish to implement a large-scale citywide slum upgrading programme. On the positive side, both cases reinforce the need to establish an institutional and organizational environment that supports local governments in engaging in multiple, comprehensive and multi-year programming and implementation efforts. Both cases highlight the vital

[36]França and Hoelzel (2011).

[37]Cities Alliance and Municipality of São Paulo (2008).

[38]The cities involved in the international policy dialogue on slum upgrading sponsored by the Cities Alliance and sponsored by the city of São Paulo were: Cairo, Ekurhuleni, Lagos, Manila, Mumbai, and the host São Paulo.

role played by the single urban project for each favela/slum comprised of the area-based plan, the urban layout design, and the street network, which materialize the street-led upgrading strategy.[39] Undoubtedly, this lays down the foundation for land tenure regularization, street addressing, provision of garbage collection, infrastructure supply, improved quality of life, improved safety, spatial connectivity, and better accessibility to the nearby districts and the city as whole.

However, this is not enough to guarantee the full inclusion of the upgraded slums (now transformed into improved neighbourhoods) into the urban management and planning systems of the city.

Ten years after the implementation of the programmes and projects, the improvement was significant and the municipalities were able to unveil significant achievements: a total of 54 favelas were upgraded during the first phase and 89 during the second phase; the impressive amount of 500 km of water pipes and 548 km of sewage pipes (more than the distance from London to Paris) were delivered; and 1.7 million^2 of roads were paved.[40] However, on the negative side, inclusion into the municipality's property registry and the priority list of service providers, as well as the tenure regularization, did not happen automatically. Considering that the land property regularization in one of the first upgraded favelas by the Favela Bairro Program took nearly 20 years to be achieved,[41] we can conclude that the area-based plan and the street-led approach alone that were adopted by the programme were not sufficient to enable the full integration of these areas into the city. Certainly, time and resources but also political will, in addition to the complexity of property regularization, were all factors. The legalization and registration of tenure in a context of densely occupied settlements are not easy tasks to conduct, since they are time-consuming processes that involve multiple steps and institutions.

Another positive aspect of both the São Paulo and Rio programmes is that they considered the integration of the spatial/physical improvements together with social, economic, institutional, and environmental actions in order to help to improve the access to jobs and income, as well as to support youth, women-headed households, health, and environmental resilience. In Rio de Janeiro, the programme also included the establishment of a POUSO,[42] the Municipal Office for Social and Urbanistic/Building Orientation, in each favela after completion of the upgrading works as a way of building capacity among residents. This was a successful initiative and, at one point, there were more than 50 POUSOs in the city. Unfortunately, the POUSOs were discontinued due to a lack of funding. This is a common fate for urban renewal and slum upgrading programmes globally, a fate that needs to be changed if we aim to achieve the transformational Agenda 2030 in cities and human settlements.

[39]Conde and Magalhães (2004).

[40]Municipality of Rio de Janeiro and InterAmerican Development Bank (2003).

[41]http://www.rio.rj.gov.br/web/guest/exibeconteudo?id=7766264.

[42]Posto de Orientação Urbanistica e Social. This is an advanced administrative office of the municipality of Rio staffed by engineers, architects, social workers, and community builders which provided technical advice as well as social support to residents. It was instrumental to implementing the land-use plan and the minimal building parameters for those willing to expand their housing units.

To conclude, institutions usually do not change their behaviour quickly towards the neighbourhoods that were once slums, and stigma still continues to hinder the social and spatial inclusion of these areas. To change this trend, community leaders and social and human rights activists have to continue pushing for this change. Leaving no one and no place behind takes more than physical and spatial inclusion, and it takes much longer than the lifespan of the projects and programmes. As one favela dweller says: “Things are getting better. But street names don’t stop people getting killed” (Scudamore 2010: 135).[43]

References

Acioly CC (1992) Settlement planning and assisted self-help housing: an approach to neighbourhood upgrading in a Sub-Saharan City, Publikatie Bureau. Delft University of Technology, Delft

Acioly CC (1993) Planejamento Urbano, Habitação e Autoconstrução: experiências com urbanização de bairros na Guiné-Bissau. Publikatieburo, Faculty of Architecture Press, Delft University of Technology

Acioly CC (2000) Possibilities to start a citywide project in Albania. A report prepared for the Netherlands Habitat Platform

Acioly CC (2001a) Reviewing urban revitalisation strategies in Rio de Janeiro: from urban project to urban management approaches. In: GeoForum: Special issue on Urban Brazil, vol 32(4). Elsevier, UK

Acioly C (2001b) Reviewing urban revitalization strategies in Rio de Janeiro: from urban project to urban management approaches. Geoforum 32:509–520. www.claudioacioly.com/downloads/articles/Acioly%202001_Revitalisation%20Strategies%20Rio_Geoforum.pdf

Acioly CC (2010) The informal city and the phenomenon of slums: the challenges of slum upgrading and slum prevention. International New Town Institute (2010) New Towns for the 21st Century. The Planned vs. the Unplanned City. SUN Architecture, Amsterdam, pp 222–231

Acioly C (2014) Street-led citywide slum upgrading. Global urban lecture. http://unhabitat.org/street-led-city-wide-slum-upgrading-claudio-acioly-un-habitat/

Acioly CC, Aliaj Band Kuçi F (2003) A path road to citizen participation in urban management. Lessons learned from an Albanian NGO. Unpublished article

Acioly CC, Davidson F (1996) Density in urban development. Lund University, SIDA and LCHS, Building Issues 3

Ahmedabad Municipal Corporation (2005) Ahmedabad slum networking program, submitted for the Dubai International Award for best practices to improve the living environment

Amis P (2001) Rethinking UK aid in urban India: reflections on an impact assessment study of slum improvement projects. Environ Urbaniz 13(1). http://eau.sagepub.com/

Baker JL (2008) Urban poverty: a global view. Urban Sector Board, The World Bank, Washington D.C

Banerjee B (2002) Mainstreaming the urban poor in Andhra Pradesh, India. In: Westendorff D, Eade D (eds) Development and cities. Oxfam GB, pp 204–225

Banerjee B (2006a) Impact of tenure and infrastructure programmes on housing conditions in squatter settlements in Indian cities. In: Shaw A (ed) Indian cities in transition. Orient Longmans, New Delhi

Banerjee B (2006b) Giving voice to the poor in municipal planning in Andhra Pradesh. In: Rao PSN (ed) Urban governance and management: Indian initiatives. IIPA, New Delhi, pp 213–235

[43] Scudamore (2010).

Banerjee B (2010) Slum poverty in Asia: characteristics, policy responses, new challenges and opportunities. Background paper for conference on the "The Environments of the Poor", 24–26 Nov 2010. Asian Development Bank, New Delhi

Banerjee B (2011) Urban transformation and inclusion of the poor in Asian cities: the case of participatory municipal planning. EWC-UPENN workshop changing cities—linking global knowledge to local action 26–28 Sept 2011, Honolulu, Hawaii

Betancur JJ (2007) Approaches to the regularization of informal settlements: the case of PRIMED in Medellin, Colombia. Global Urban Dev Mag 3(1)

Booth C (1889) Life and labour of the people of London. http://www.umich.edu/~risotto/partialzooms/ne/50nek56.html. Accessed on 15/02/2012)

Brakarz J et al (2002) Cities for all. Recent experiences with neighbourhood upgrading programs. Inter-American Development Bank, Washington D.C.

Buckley R, Kalarickal J (eds) (2006) Thirty years of World Bank shelter lending. What have we learned? World Bank, Washington

Cities alliance (2002a) Citywide slum upgrading. http://www.citiesalliance.org/ca/sites/citiesalliance.org/files/Anual_Reports/upgrading_9.pdf

Cities Alliance (2012b) A citywide slum upgrading programme and CDS in Agra, India. http://www.citiesalliance.org/ca/node/2696. Accessed on 5/2/2012

Cities Alliance and Municipality of São Paulo (2008) Slum upgrading up close. Experiences of six cities. An international policy dialogue: challenges of slum upgrading in São Paulo, Brazil. P110791. Cities Alliance, Washington

Cities Alliance (undated) Cities Alliance in Action. Technology that transformed urban planning in São Paulo: HABISP. Cities Alliance, Washington

Citynet (1995) Municipal land management in Asia: a comparative study. United Nations,New York

Clarke J, Jones S, Pickford J et al (1989) Evaluation study of Hyderabad slum improvement project. DFID London, Evaluation report EV 475

CODI (2007) 80 Community development projects

COHRE (2008) Successes and strategies: responses to forced evictions. Centre on Housing Rights and Evictions, Geneva

Conde LP, Magalhães S (2004) Favela-Bairro: rewriting the history of Rio (Favela Bairro: uma outra história da cidade do Rio de Janei). Vi, verCidades, Rio de Janeiro

Cuenin F (2010) Economic analysis for settlement upgrading programmes. In: Rojas E (ed) Building cities. Neighbourhood upgrading and urban quality of life. Inter-American Development Bank, Cities Alliance, David Rockefeller Center for Latin American Studies, Harvard University, Washington

Czaenowsky TV (1986) The streets as communications artifact. In: Anderson S (ed) On streets. The MIT Press, Cambridge

Delhi Development Authority (DDA) (2007) Delhi Master Plan 2021. Rupa Publishers, New Delhi

Denaldi R (1997) Viable self management: the FUNACOM housing programme of the São Paulo municipality. Habitat International 21(2):213–227

Development Innovations Group (2006) Best practices in slum improvement: the case of Ahmedabad India. http://www.housingfinanceforthepoor.com/

Digital Mapping to Put Slums on the Map (2009) http://www.nowpublic.com/tech-biz/digital-mapping-put-slums-map. Accessed on 20/03/2012

Do Prado Valadares L (2005) A Invenção da Favela. Do mito de origem a favela.com. Editora FGV, Rio de Janeiro

Durand-Lasserve A (1998) Law and urban change in developing countries: trends and issues. In: Fernandes E, Varley A (eds) Illegal cities: law and urban change in developing countries. Zed Books Ltd., London

Durand-Lasserve A, Clerc V (1996) Regularization and integration of irregular settlements: lessons from experience. UMP Working Paper Series no. 6. UNDP/UNCHS-HABITAT/World Bank

Farvacque-Vitkovic C, Godin L, Leroux H, Verdet F, Chavez R (2005) Street addressing and the management of cities. World Bank, Washington

Favro C, Soll I (eds) (1994) Streets, critical perspectives on public space. University of California Press, USA

Fernandes E (2011) Regularization of informal settlements in Latin America. Policy focus report. Lincoln Institute of Land Policy, Cambridge

França E (2013) Slum upgrading: a challenge as big as the City of São Paulo. Focus 10(1), Article 20. https://doi.org/10.15368/focus.2013v10n1.10. Available at: http://digitalcommons.calpoly.edu/focus/vol10/iss1/20

França E, Hoelzel F (2011) Integrating informal cities: prime challenge for megacities of the south, 14 Apr 2011. https://www.boell.de/en/navigation/urban-development-elisabete-franca-fabienne-hoelzel-sao-paulo-11732.html

Gateway to Urban Planning and Management in Asia (1996) The Orangi Pilot Project in Karachi Pakistan. The best practices initiatives human settlements in Asia. http://www.hsd.ait.ac.th/bestprac/orangi.htm

Gehl J (1996) Life between buildings. Danish Architectural Press, Copenhagen

Gehl J, Gemzoe L (2006) New city spaces. The Danish Architectural Press, Copenhagen. Ghousia Colony case study. www.globenet.org/preceup/pages/ang/chapitre/capitali/cas/pakist_b.htm. Accessed on 3/3/2012

Ghel J, Gemzoe L, Kirkneas Sodergaard (2006) New city life. Danish Architectural Press, Copenhagen

GLTN (2008) Enumeration as a grassroot tool towards securing tenure in slums: an initial assessment of the Kisumu experience. A GLTN working paper

Government of Andhra Pradesh (2004) MAPP guidelines. Directorate of Municipal Administration

Government of India (2011) Rajiv Awas Yojana: guidelines for slum-free planning. Ministry of Housing and Urban Poverty Alleviation, Government of India. http://mhupa.gov.in/w_new/RAY%20Guidelines-%20English.pdf

Great Britain National Institute if Urban Affairs (2011) Reforming agra by re-imagining through slum up-gradation. http://www.indiaurbanportal.in/Bestpractices/Bestpractices103/Bestpractices103589.PDF. Accessed on 5/2/2012

Gulyani S (2010) The living conditions diamond: an analytical and theoretical framework for understanding slums. Environ Plann A 42:2201–2219

UN HABITAT (2008) Secure land tenure for all. Global Land Tool Network

Hamdi N, Goethert R (1997) Action planning for cities: a guide to community practice. Wiley, New York

Handzic K (2010) Is legalized land tenure necessary in slum upgrading? Learning from Rio's land tenure policies in the Favela Bairro Program. Habitat Int 34(2010):11–17

Hardoy J, Satterthwaite D (1993) Housing policies: a review of changing government attitudes and responses to city housing policies in the Third World. In: Cheema GS (ed) Urban management policies and innovations in developing countries. Greenwood Praeger Press, Westport, pp 111–160

Hasan A (2006) Orangi Pilot project: the expansion of work beyond Orangi and the mapping of informal settlements and infrastructure. Environ Urban 18(2):451–480

Herrmann M, Svarin D (2009) Environmental pressures and rural-urban migration: the case of Bangladesh. UNCTAD, Online at http://mpra.ub.uni-muenchen.de/12879/ MPRA Paper No. 12879

Homeless International (2005) Community-led infrastructure finance facility—3rd annual review, London

ICICI Property Services (2009) Dharavi a realty story 2025, 20 July 2009, download from Facebook

IDB (2011a) 2010 Development effectiveness report. http://www.iadb.org/en/annual-meeting/2011/annual-meeting-article,2836.html?amarticleid=9167

IDB (2011b) Improving living conditions in low-income neighborhoods in Rio de Janeiro. http://www.iadb.org/en/annual-meeting/2011/annual-meeting-article,2836.html?amarticleid=9164

Imparato I, Ruster J (2003) Slum upgrading and participation: lessons from Latin America. World Bank

Infrastructure in India (2006) Infrastructure report 2006, 3i Network. Oxford University Press, New Delhi
Jacobs J (1961) The death and life of Great American Cities. Vintage Books, New York
Jacobs AB (1993) Great streets. Massachusetts Institute of Technology, USA
Kar K et al (1997) Participatory impact assessment of Calcutta slum improvement project: main findings report. Calcutta Metropolitan Development Authority
Khosla R (2011) Inclusive neighbourhoods make inclusive cities: design and planning of slums in Agra. National Seminar on Design and Planning for Sustainable Habitat, 15–16 July, New Delhi. IDRC and Government of India, Ministry of Housing and Urban Poverty Alleviation
Local Government and Community Action (2003) Pathways to responsive and sustainable environmental improvement for the urban poor: the example of Andhra Pradesh. In Ghosh A (2003) Urban environment management—concept, New Delhi, pp 189–215
Magalhães F (ed) (2016) Slum Upgrading and Housing in Latin America. Inter American Development Bank, Washington
Martim R (1983) Upgrading. In: Skinner RJ, Rodell MJ (eds) People poverty and shelter. Problems of self-help housing in the third world. Methuen & Co. Ltd., London
McKinsey Global Institute (2011) Urban world: mapping the economic power of cities. Ministère de l'Écologie, du Développement et de l'Aménagement durables, Emerging Cities. www.villesendevenir.com
Ministry of Cities Brazil (2010) Habitar Brazil
Mossop E (2004) Opening the city, pp 41–50. In: Gastil RW, Ryan Z (eds) Open new designs for public space. Van Allen Institute, New York
Moughtin C (1992) Urban design: street and square. Butterworth Architecture
Municipality of Rio de Janeiro and InterAmerican Development Bank (2003) Favela Bairro. Ten years integrating into the city. Fundação Universitária José Bonifacio, Rio de Janeiro
Narayan D, Shah T (2000) Connecting the local to the global: voices of the poor. Framework paper prepared for 11–13 December 2000 Workshop on Local to Global Connectivity for Voices of the Poor, World Bank, Washington, D.C. http://siteresources.worldbank.org/INTPOVERTY/Resources/335642-1124115102975/1555199-1124741378410/dec00_narsha.pdf. Accessed 13 July 2011
Neighborhood Upgrading and Shelter Sector Project (2008) http://www.adb.org/publications/neighborhood-upgrading-and-shelter-indonesia
Parikh HH (1995) Slum networking: a community based sanitation and environment programme, experiences of Indore, Baroda and Ahmedabad
Participatory Development Project (PDP) (2011) Maximising use value: action guide for informal areas, GIZ
Pasteur D (1979) The management of squatter upgrading. Saxon House, United Kingdom
Prefeitura de São Paulo (2008a) Urbanização de Favelas. A experiência de São Paulo. Boldarini Arquitetura e Urbanismo, São Paulo
Prefeitura de São Paulo (2008b) Habitação de Interesse Social em São Paulo: desafios e novos instrumentos de gestão. Cities Alliance and Prefeitura de São Paulo, São Paulo
Prefeitura do Rio de Janeiro (1995) Política Habitacional da Cidade do Rio de Janeiro
Prefeitura do Rio de Janeiro (2004) From removal to the urban cell. The urban social development of Rio de Janeiro Slums
Prefeitura do Rio de Janeiro (Undated) Favela-Bairro; 10 anos integrando a cidade
Rapoport A (1987) Pedestrian street use: culture and perception. In: Moudon AV (eds) Public streets for public use. Van Nostrand Reinhold Company Inc., New York
Richartz R (1988) Improvement of an inner-city basti of Karachi: the second evaluation survey in Ghousia Colony. Free University, Amsterdam, 132 pp. Urban Research Working Papers, No. 17
Rodrigues E, Barbosa BR (2010) Popular movements and the city statute in the city statute of Brazil: a commentary. Cities Alliance

Rojas E (2010) Building cities. Neighbourhood upgrading and urban quality of life. Inter-American Development Bank, Cities Alliance, David Rockefeller Center for Latin American Studies, Harvard University, Washington

Royal University College of Fine Arts (2008) Dharavi. Documenting Informalities. Royal University College of Fine Arts, Stockholm

Scudamore J (2010) Heliopolis. Vintage Books, London

Shehayeb D (2009) Advantages of living in informal areas. In Kipper R, Fischer M (eds) Cairo's informal areas between urban challenges and hidden potentials: facts, voices, visions. Cairo, Egypt, GIZ, pp 35– 43. http://www2.gtz.de/dokumente/bib/gtz2009-0424en-cairo-informal-areas.pdf

Shehayeb D, Sabry S (2008) Living in informal areas: advantages and disadvantages-Shiakhat Boulaq Al-Dakrour. An "unpublished" background study for GIZ towards a presentation in the International Seminar "Exchanging Global and Egyptian Experiences in Dealing with Informal Areas within the Wider Urban Management Context", 14–15 Oct 2008 organized by GIZ, Cairo, Egypt

Sindh Katchi Abadis Authority (2012) Presentation on Katchi Abadis in Karachi. www.urckarachi.org/SKAA%20Presentation.pps. Accessed on 3/3/2012

Skinner T, Wegelin EA (eds) (1987) Shelter upgrading for the urban poor: evaluation of third world experience. Island Publishing House, Manilla, Philippines

Slum Rehabilitation Authority (2004) Dharavi redevelopment plan. Govt. of Maharashtra, Housing and Special Assistance Department. www.dharavi.org

Society for Promotion of Area Resource Centres (SPARC) and Kamala Raheja Vidyanidhi Institute of Architecture (KRIVA) (2010) Reinterpreting, Reimagining, Redeveloping Dharavi, Mumbai

SPARC (1985) We the invisible: survey of pavement dwellers of Mumbai

ThinkSoft Consultants (2005) Social impact study of Andhra Pradesh Urban Services for the Poor (APUSP)

Toutain O, Gopiprasad S (2006) Planning for urban

UN Millennium Project (2005) A home in the city. Task force improving lives of slum dwellers. Earthscan, London

UNCHS (1980) Physical improvement of slums and squatter settlements. Report of an Ad Hoc Expert Group Meeting, Nassau, 31 Jan.-4 Febr. 1977 Nairobi

UN-ESCAP (No Date) Local Government in Asia and the Pacific: a comparative analysis. www.unescap.org/huset/lgstudy/index.htm

UN-HABITAT (2003) The challenge of slums. Global Report on Human Settlements 2003. Earthscan and UN-HABITAT. Earthscan, London

UN-HABITAT (2006a) Analytical perspective of pro-poor slum upgrading frameworks. UN-HABITAT, Nairobi

UN-HABITAT (2006b) State of the world's cities 2006/2007, UK, Earthscan

UN-HABITAT (2008a) State of the world cities Report 2008–2009

UN-HABITAT (2008b) Women's safety audits: what works and where? UN-Habitat, Nairobi

UN-HABITAT (2009) Assessment of safety and security issues in slum upgrading initiatives: the case of Favela-Bairro, Rio de Janeiro. Draft version. Safer Cities Programme

UN-HABITAT (2010a) Citywide action plan for upgrading unplanned and unserviced settlements in Dar es Salaam. UN-HABITAT, Nairobi

UN-HABITAT (2010b) Count me in. Surveying for tenure security and urban land management. United Nations Human Settlements and Global Land Tool Network, Nairobi

UN-HABITAT (2010c) State of the World's cities report 2010–2011. Bridging the Urban Divide

UN-Habitat (2010d) State of the world's cities 2010/2011-Cities for all: bridging the urban divide. UN-Habitat, Nairobi. Earthscan, London

UN-HABITAT (2011) Building safety through slum upgrading. UN-HABITAT, Nairobi

UN-Habitat (2012a) Streets as tools for urban transformation in slums: a street-led approach to citywide slum upgrading. UN-Habitat, Nairobi

UN-Habitat (2012b) State of the world's cities 2012–2013—prosperity of cities. UN-Habitat, Nairobi. Earthscan, London

UN-Habitat (2013a) Streets as public spaces and drivers of urban prosperity. UN-Habitat, Nairobi

UN-Habitat (2013b) The relevance of street patterns and public space in urban areas. UN-Habitat Working Paper April 2013. UN-Habitat, Nairobi

UN-Habitat (2016) World cities report 2016. Urbanization and development: emerging futures. UN-Habitat, Nairobi

UN-Habitat (2017) Brazil. Impact Story. Nairobi: Participatory Slum Upgrading-PSUP. UN- Habitat

UN-HABITAT and OHCHR (2010) Fact Sheet 21: The Right to Adequate Housing

UN-HABITAT & UNESCAP (2008) Housing the poor in Asian cities; No. 2 Low Income Housing, Quick guides for policy makers

University of Birmingham, Development Administration Group (1998) Report of the impact assessment study of slum improvement projects in India, DFID

Van Horen B (2000) Informal settlement upgrading: bridging the gap between the de facto and the de jure. J Plann Educ Res Summer 2000(19):389–400

Varley A (2002) Private or public: debating the meaning of tenure legislation. Int J Urban Reg Res 26(3)

Viloria-Williams (2006) Urban community upgrading, lessons from the past – prospects for the future. World bank Institute

World Bank (1993) Housing: enabling markets to work. Washingto D.C., World Bank

World Bank (1995) Indonesia impact evaluation report enhancing the quality of life in urban Indonesia: the legacy of Kampung Improvement Program, Report No. 14747-IND. Operations Evaluation Department, World Bank. http://go.worldbank.org/Q4SEO02CB1

World Bank-UNDP Water and Sanitation Program South (WSP) Asia (1998) Ahmedabad Parivartan. http://www.rio.rj.gov.br/web/guest/exibeconteudo?id=7766264

Resilient Slum Upgrading in Indonesia. Improving Spatial Connections from Bottom and Top

Oscar Carracedo García-Villalba

Abstract Traditional kampung forms a large part of the urban real-estate in Indonesia. These kampung, home to large numbers of low-income residents, suffer from environmental problems such as a lack of basic services and infrastructure (Ernawati et al. in Human Soc Sci 1(1):1–6, 2013). This chapter studies the Kampung Improvement Programme (KIP) as the first of its kind, implemented by the Indonesian government. It stands as the most successful pioneer slum upgrading program based on infrastructure provision ever undertaken. As mentioned by Devas (Ekistics, No. 286, 1981: 19) "the Kampung Improvement Programme in Indonesia is often quoted as an example of a successful approach to the housing needs of the urban poor. It certainly has been impressive in its scale—improving living conditions for something like half the population of the city of Jakarta". The success of the KIP is demonstrated through an analysis of the case study of the award-winning project for Kampung Kebalen as part of the Surabaya Comprehensive KIP.

Keyword Informal settlements · Slums · Kampung Improvement Programme · Community-based urban regeneration · Spatial connectivity · Social resilience · Indonesia

1 Introduction

Informal settlements with a lack of basic infrastructure and public services (water, sanitation, electricity) and substandard housing are becoming the norm in the urban environments of the Global South economies. However, after decades of aggressive and violent approaches intended to eradicate these neglected areas, on-site slum upgrading programs to improve the living conditions of these settlements are now on the rise.

O. Carracedo García-Villalba (✉)
Director Master of Urban Design, Director Designing Resilience in Asia International Research Programme, Department of Architecture, School of Design and Environment, National University of Singapore, Singapore, Singapore
e-mail: oscar_carracedo@nus.edu.sg; omc@coac.net

O. Carracedo García-Villalba (ed.), *Resilient Urban Regeneration in Informal Settlements in the Tropics*, Advances in 21st Century Human Settlements,
https://doi.org/10.1007/978-981-13-7307-7_3

The Kampung Improvement Programme (KIP) in Indonesia is, according to UN-Habitat (2006), the world's first urban slum upgrading project. Initiated in 1969 in the capital of Indonesia, Jakarta, the programme proposed an innovative approach through the provision and improvement of basic urban services, such as roads and footpaths, water supply, drainage and sanitation, public toilets, solid waste disposal, as well as health and education facilities. In other words, the main objective of the programme is to enhance the neighbourhood living conditions, improve the sanitation facilities and the environmental conditions, and to uplift a low-income sector of the population.

Kampung are the main urban settlement type that houses the majority of Indonesia's inhabitants. They form a fundamental part of the urban structure of cities in Indonesia. Historically, kampung were autonomous settlements, villages located on pockets of rural land on the city fringes that, due to rapid urbanisation and immigration flows, became subject to overcrowding, poor sanitation standards and a prevalence of poverty (Djajadiningrat 1994). Kampung are informal, unplanned and unserviced housing areas (Devas 1981). This distinctive housing type is the only affordable housing option for both long-term residents and newcomers seeking the benefits and services provided in the city, such as education, employment, healthcare and amenities. Despite its functional importance and deep historical roots in South-east Asian urbanism (Hawken 2017), the kampung settlement has been considered a slum-like habitat due to the lack of basic urban services. Scholars frequently mention the role of kampung in accommodating the city's inhabitants, however, the settlements are often described as transitory with pathological elements inherent within the type (Shirleyana et al. 2018). In this sense, kampung are usually over-congested areas located in strategic parts of the city. They are generally located surrounding wealthy neighbourhoods and near economic areas that kampung residents can take advantage of.

2 The Origins of the Kampung Improvement Programme: Initiatives, Objectives and Components

Historically, the initiatives related to kampung improvement were first introduced under Dutch colonial rule in 1918, before the country became officially independent in 1945. Under the name of "Kampung Verbetering", "Kampung improvement" in Dutch (Atman 1975), the first initiative was a product of the members of the opposition in the Dutch Parliament during the colonial government, who demanded better living and 'humane' conditions for local populations residing in the urban areas in the colonies. With the establishment of the municipal governments early in the twentieth century, kampung improvement became a renewed topic of interest that extended from the early 1920s to the beginning of World War II (World Bank 1995). It was in 1924 when the municipal governments of Surabaya and Semarang initiated the improvement of some of the cities' kampung. However, after Indonesia gained independence, kampung improvement diminished for a period of 20 years, due to the difficulty of sustaining the initiative politically and economically (Devas 1981; Djajadiningrat 1994). It is important to note that the political attitudes in addressing the issues with the kampung have been very different from approaches to informal settlements in other places in the world. This is due to the economic and social costs of implementing urban renewal schemes that involve mass demolitions and the interim relocation of residents during the construction of new housing. In the case of Indonesia, the innovative approach was that the government recognised the kampung as permanent settlements and adapted the "site and services" approach to existing informal settlements, positing the existing building stock as an asset to maintain, support and enhance, while preserving the local identity.

The first official Kampung Improvement Programme (KIP) was initiated by the government of Indonesia and the city of Jakarta in 1969. Under the first National Development Plan, Repelita I.[1] This plan aimed to upgrade the infrastructure and services of some 24 km^2 of city kampung within the five-year period from 1969 to 1974. As mentioned by Subagio (1986), the emphasis of the KIP was to implement specific actions to provide the basic physical infrastructures and service standards in relation to a package of eight main components: (1) upgrading and improving vehicular roads, with associated drains; (2) upgrading and paving footpaths; (3) rehabilitating and creating kampung-wide drainage; (4) providing garbage bins and collection vehicles; (5) providing safe drinking water through public taps; (6) constructing public washing and toilet facilities for clusters of kampung; (7) building neighbourhood health clinics; and, finally, (8) constructing primary school buildings.

The Mohammed Husni Thamrin Programme (MHT-KIP) was the first kampung improvement project implemented in Jakarta. The programme had four specific objectives: building the basic requirements for the city's population, such as roads, footpaths, drainage, sanitation, drinking water supplies, primary school buildings and health clinics; identifying the poorest part of the population suffering from inadequate environmental conditions; drawing up an investment programme to benefit the broadest section of the population; and encouraging the population to realise its potential for self-sufficiency and cooperation within their kampung (KIP Unit DKI Jakarta 1991). Due to the success of this first pilot project, the KIP was extended throughout the city. From the period 1969–1984, the KIP affected 5 million beneficiaries and 25% of the city (World Bank 1995).

Also in 1969, and in parallel to this case, the city of Surabaya developed the Wage Rudolf Supratman Programme (WRS-KIP). Surabaya is the second largest city after Jakarta, with more than 3 million inhabitants (Municipal Government of Surabaya 2015) and large parts of the city are covered with kampung, providing housing options especially for low-income households. It is estimated that more than 60% of the city's inhabitants live in these areas (Duncan 2004). In contrast with the case of Jakarta,

[1]Indonesian national development is characterised by Long-Term Development (PJP: Pembangunan Jangka Panjang) and Five-Year Development Plans (Repelita: Rencana Peinbangunan Lima Tahunan). The former sets broader goals and objectives for national development over 25 years, while the latter sets up detailed sector policies and targets. Repelita is the operational plan to achieve long-term goals and objectives (Kuswardono 1997).

and recognising its uniqueness, the W.R. Supratman project emphasised community self-help, both in taking the initiative for planning and implementation, as well as encouraging the community to actively construct the infrastructure in the kampung (Djajadiningrat 1994). The public works programme began as a physical infrastructure programme, like in Jakarta. However, soon it was realised that, to be successful, significant community involvement was needed due to the limited financial and technical resources in Surabaya. As Silas (1992) highlights, the attractiveness of this approach was that community contributions were matched by government funds, typically up to 50% of the required budget.

Realizing the success of the KIP in Jakarta and Surabaya, the national government expanded the KIP programmes to other cities in Indonesia, encompassing 50,000 ha and 15 million beneficiaries (World Bank 1995). In the late 1990s, the KIP had been implemented in almost 800 cities across Indonesia and this massive impact, as mentioned by Harari and Wong (2017), led practitioners and policymakers to view the programme as an example of a successful approach to slum upgrading.

3 KIP, from Institutional to Community-Based Development. Balancing Planning, Community and Funding

As mentioned before, one of the essential aspects of the KIP is that the local government recognised the kampung as permanent settlements from the very beginning of the planning process. Being aware of the limited funds, the large areas occupied, and the population living in kampung, the planning of the KIP could only be guided by the principle of on-site upgrading as the only feasible and realistic management option without changing the existing social structure. Therefore, due to the limited resources, the first step was to meet the population's the basic needs, providing the spatial support infrastructure from a city-wide perspective, and later, in a second stage, to reinforce the kampung with health and educational facilities.

The first steps of the KIP in Jakarta under Repelita I started as a local government responsibility, administered by various departments with poor coordination. This lack of organisation resulted in an essential change in the program implementation under Repelita II, when the planning process was recentralised with the creation of a special project unit responsible for planning and implementing: the BAPPEM-PMHT (Badon Pelaksana Pembangunan—Proyek Mohammad Hum Thamrin). This unit was organised in a very hierarchical way, but the responsibilities of the different actors were very clearly defined (Devas 1981). Except for the case of Surabaya, in both the Repelita I and II (1969–1979) the decisions and the initiatives to develop the program were taken by the government and the unit, which decided on their own about the different planning, design, implementation and finance components, showing that the MHT-KIP was clearly characterised by a top-down approach to solving the physical problems related to the poor living conditions in the kampung.

It was not until the Repelita III (1979–1984) that the government introduced changes to the planning system to address some of the concerns about integrating community participation into the coordination, finance and development of the programme. These changes turned the KIP into a combination of both "top-down" and "bottom-up" systems, with the kampung communities participating in the programme, which had some specific implications. On the one hand, the programme was still an initiative of the local government and the planning unit but, on the other hand, the kampung communities were able to develop their own proposals aimed to improve physical, social and economic aspects. The detailed community proposals had to be agreed upon through community consultation and discussion in the LKMD, the Organisation for Community Security (Lembaga Ketahanan Masyarakat Desa), and after that submitted to the BAPPEM PMHT for approval and budgeting (Djajadiningrat 1994).

We can say that the W.R. Supratman programme somehow inspired this change in the KIP planning process in Surabaya. Although in this case, the approach was similar to Jakarta, the much limited financial and technical resources in the municipality of Surabaya meant that the programme could only be realised successfully with extensive community involvement. Therefore, since the very beginning, the government in Surabaya had to target the efficient mobilisation of the communities to improve and manage their own living environments.

The successful implementation process of the KIP and its impact would not have been possible without funding from the national (APBN, Anggaran Pendapatan dan Belanja Negara), provincial (APBD I, Anggaran Pendapatan dan Belanja Daerah I) and city governments (APBD II, Anggaran Pendapatan dan Belanja Daerah II) (Sarkawi 2015), but especially without the financial support of the World Bank, which facilitated the extension of the programme to other cities in the country over a period of 14 years (1974–1988). The World Bank entered the KIP in Jakarta in 1974 with the Urban I project, which provided a loan of US$25 million for the period 1974–1976 to establish a national urban development programme that would improve the living conditions of the urban poor by increasing their access to better physical infrastructure and housing (World Bank 1995). As a result, more than 2000 ha and 890,000 people benefited from the improvement of spatial connectivity.

In the Urban II, the second project for the 1976–1979 period, the bank provided a loan of US$43 million to the KIP in Jakarta and also, for the first time, to the KIP in Surabaya. In the latest case, as part of the strategic bottom-up approach for the city, communities were responsible for providing the land and the space required for the new layout of the spatial connectivity, and they were also responsible for reorganising the on-site relocation of the people affected by the project. In addition, the community was responsible for organising and operating the maintenance of the facilities provided and paid for from the community funds. This strategic approach rooted in community involvement made the residents feel like part of the programme and enhanced the sense of belonging, which ensured their commitment to the maintenance of the infrastructure provided.

The benefits that this new slum upgrading approach of the KIP brought to 1,636,000 people were enough of an incentive for the World Bank to fund two more projects: the Urban III from 1979 to 1981, with a loan of US$54 million for projects in Jakarta, Surabaya, Ujung Pandang, Semarang and Surakarta; and, finally, the Urban IV, for the period from 1981 until 1988, with the provision of a loan of US$43 million to cover projects in Banjarmasin, Denpasar, Padang, Palembang, Pontianak and Saramarinda. In summary, the four Urban projects benefited a total of 4,292,250 people, improving the spatial infrastructural connectivity of 11,331 ha (World Bank 1995).

As mentioned before, the primary upgrading criteria for the World Bank to support the KIPs involved spatial connectivity and infrastructure provision to integrate the kampung into the rest of the city. However, it is fundamental to understand how the institution took a clear position to facilitate the involvement of community-based organisations in the implementation programmes, understanding that communities play a strategic role and constitute the structural backbone for the success of the KIP. In this sense, and despite the concerns about how the projects might achieve a cost recovery, one of the fundamental policies in the ideology behind all the Urban projects funded by the World Bank, was that kampung residents would not be charged in any case for the infrastructure provided. Instead, all the costs for upgrading the physical infrastructure of the kampung would be recovered indirectly from citywide property taxes. However, in exchange for this policy, to promote commitment from the community and to avoid giving the wrong impression that the projects were a "gift", the residents and/or the government were required to contribute the land that was needed to implement the new access streets, street widenings, drains, footpaths or any other infrastructures without any kind of compensation except for the provision of the land needed to build the education or health facilities.

We can affirm that the improvement of millions of people lives demonstrates that the KIPs have achieved their physical objectives of integrating the kampung into the city. Still, at the same time, as cited by The World Bank (1995), they have failed at cost recovery. However, the improvements in infrastructure and integration with the city, the enhanced household economies as a result of that integration, the provision of basic human needs, and poverty alleviation might be enough reasons to justify the failure in achieving the objective of cost recovery.

Summarizing, we can state that the KIPs success and substantial impact on the lives of millions of people is mostly a consequence of the delicate and intricate balance of reasonably efficient implementation through simple but strongly institutionalised and centralised planning processes, backed by a very strong ground-up involvement and community participation, and of course the essential sustained, but also permissive, funding support from the World Bank.

4 Resilience in the Focus of the KIP. Spatial Connectivity to Achieve Flooding Resilience

As explained in earlier sections, the main focus of the KIP was on the upgrading and improvement of spatial connectivity through street layout, water infrastructure and footpaths. I would like to suggest that the KIP approach also pioneers the concept of resilience. In this case, the incorporation of a new perspective on the meaning of upgrading informal settlements through spatial connections refers more specifically to social resilience and flood-resistant infrastructure.

Flooding is one of the most critical environmental hazards affecting kampung, and it is mostly associated with drainage problems due to a lack of maintenance or illegal dumping of garbage. Flooding poses several environmental and health risks due the pollution of water by human waste or because of waterlogging and the insect breeding effect. As mentioned in the World Bank report (1995), drainage systems were often not connected to the broader infrastructure, causing backlogs at the entrance of the city-wide drainage systems. According to the evaluation made by Silas (1983), 48.8% of the kampung in the KIP were affected by flooding. However, the interest in flooding and the funding for its solution did not take on special relevance until the Urban III project, when the improvement of drainage systems and waste management were officially incorporated. At that moment, flooding was included as one of the essential criteria for a kampung to receive the improvement programme support (Sarkawi 2015). There were two main criteria that the selected kampung were meant to follow: a primary criteria concerning problems of water supply, flooding, sanitation and health facilities; and a second criteria related to accessibility within the kampung, population density, average income level of the population, educational facilities, and age of the kampung (SKIP 1981).

4.1 The Case of Kampung Kebalen in Surabaya

Elaborating on the case of kampung Kebalen in Surabaya as a paradigmatic example of an innovative perspective on the meaning of upgrading informal settlements, this section suggests that the project for the kampung incorporates different measures to achieve flood resistance and social resilience.

The KIP project for Kampung Kebalen is located in the city centre of Surabaya, close to the Kalimas River and the Kalimas train station, and not far from the harbour. At the moment of the realisation of the KIP project, the kampung was inhabited by approximately 60,000 people living in 20,000 houses distributed across 32 ha. Kampung Kebalen was chosen as part of the KIP Surabaya programme since it met both the primary and secondary criteria mentioned above: an old kampung area; highly populated, with over 800 people per hectare; a low- income population, with an average household income of US$50; and with poor environmental conditions, particularly flooding that used to affect the streets during the rainy season, hampering displacement and infiltrating into the houses. The project for Kampung Kebalen started in 1976 with the programme development and the preliminary studies, followed by the design process in the year 1979, and finally the implementation in 1980. The project was finally completed in 1981. Due to the limited availability of staff within the municipal government of Surabaya, the project for Kampung Kebalen was achieved thanks to the joint efforts of the municipality, the technical assistance of professors and students from the Faculty of Architecture at the Institute of Technology Sepuluh (ITS), who helped with the preparatory process and surveys, and the close collaboration with the kampung residents through the community council, who helped to establish the needs, deficiencies and priorities of the community (SKIP 1981).

The programme aimed to propose an inexpensive method, using cheap standardised components and simple implementation procedures, to rapidly provide a basic infrastructure using minimal technical and administrative resources. According to Silas in the report prepared for the Aga Khan award (1981), potable water, electricity and sanitation were lacking, and flooding of the streets and houses during rainy seasons exacerbated the problems. Therefore, it was of vital importance to construct spatial connections through footpaths and streets incorporating

pavements and gutters as the drainage system to reduce the environmental problems associated with accessibility and flooding and to help to alleviate the deficient living standards endemic to the kampung. In addition to solving the flooding problem, the street and infrastructure improvement also had multiple effects. On the one hand, the neighbours of the kampung created a micro-climate by planting trees, flowers and shrubs in the exterior spaces of their houses opening onto the streets, and, on the other hand, the climate of the area surrounding the kampung was improved (SKIP, 1981), reducing the heat island effect.

In general, after the implementation of the KIP programme throughout Indonesia, and as a result of the improved spatial connections with the addition of drainage systems, the impact of flooding during rainy episodes has been reduced. The percentage of kampung affected by flooding decreased from 48.8 to 12.2%, demonstrating the effectiveness of the flooding resilience measures implemented through the incorporation of drainage systems (Silas 1983).

It is also interesting to highlight that some components of the Urban projects were not only designed to deal with environmental issues such as drainage and flooding, they were also intended to have an impact at the city-wide level, constituting a pioneering approach to the street-led city-wide slum upgrading projects.

5 KIP. Recognising the Success of the Innovative Resilient Upgrading Approach

The success of the innovative resilient upgrading approach of the KIP has been widely recognised internationally. The city of Surabaya was awarded the Aga Khan Award for Architecture (AKAA) in 1983 for the successful approach and physical and social improvements achieved by the KIP for Kampung Kebalen. In 1991, it also received the World Habitat Award from UN-Habitat, then the United Nations Conference on Human Settlements (UNCHS) and, in 2006, the Changemakers competition award on "How to Provide Affordable Housing," organised by ASHOKA and Habitat for Humanity (Das 2006). Finally, and more recently, in 2018, Surabaya received the Lee Kuan Yew World City Prize[2] Special Mention "for taking a bold urban development strategy to preserve and develop its kampung neighbourhoods". The Jury citation also highlighted that "the involvement and commitment of both the people and the Mayor working closely together for a better quality of life, is clearly felt through the palpable vibrancy of the kampung. Surabaya has distinguished itself as a forward-looking emerging city and is an inspiration to other cities in developing economies who are looking to learn from a well-managed city now characterised by economic growth, social harmony, and environmental sustainability."[3]

The Kebalen project in Surabaya is not the only recognition that the Kampung Improvement Programme received. The Lee Kuan Yew World City Prize also recognises the value of the general achievements of the KIP for having "successfully brought together strong community support and participation from the citizens to collaborate closely with the local government in transforming the kampung into clean, conducive, and productive environments. The initiative is also an inspiring model for alleviating poverty, through which the city provided professional training to improve the villagers' in-house production of food and crafts for sale, made available cheap credit by the national government, and prepared the market to absorb the products." Moreover, the KIP in Jakarta also received the AKAA in 1980, in recognition of its widespread impact on the lives of many people by significantly improving their environment, and for having traced a replicable and cost-effective way to deal with the massive problem of informal settlements (SKIP 1981).

After almost 50 years, the KIP is still considered one of the most impactful slum upgrading programs and, as such, it has been replicated in Vietnam, Cambodia, Thailand and India (Dhakal 2003) to uplift the low-income sector of society and provide them with better living conditions. However, a detailed look into decades of KIP projects offers some interesting insights about the limitations of the programme. Adaptation to the new population requirements and societal demands were needed. Therefore, a new approach was required to strengthen the potential of community-based approaches and to incorporate the social and economic dynamics.

---

[2]The prize is a biennial international award that honours outstanding achievements and contributions to the creation of liveable, vibrant and sustainable urban communities around the world. The Prize is awarded to cities and recognises their key leaders and organisations for displaying foresight, good governance and innovation in tackling the many urban challenges faced, to bring about social, economic and environmental benefits in a holistic way to their communities. https://www.leekuanyewworldcityprize.com.sg/about/about-the-prize.

[3]https://www.leekuanyewworldcityprize.com.sg/laureates/special-mentions/2018/2018-special-mentions/surabaya.

6 Updating the Spatial Connectivity Upgrading Approach Through Community Participation and Social Resilience

The social and community resilience dimension of the KIP can be considered one of the innovative strengths that has developed further in the updates of the programme. As defined by Magis (2010), social or community resilience encompasses "the existence, development, and engagement of community resources by community members to thrive in an environment characterised by change, uncertainty, unpredictability, and surprise."

In the case of Kampung Kebalen, close cooperation between the beneficiaries and the municipality was sustained throughout the entire process and at all levels of progression of the improvement programme (SKIP 1981). The involvement and commitment of the community in the process of conceptualising and designing the project, and later its involvement in the construction and maintenance phases through community-based initiatives, created a very resilient approach through social attachment.

Capitalising on the existing kampung social ties, the organisational capacity of the community and the gotong royong, or mutual-help/aid cooperation tradition, as fundamental assets, the project for Kebalen established three main objectives. The objectives included increasing the coexistence of social life and community collaboration; stimulating the local community to improve housing and living conditions through self-help approaches; encouraging the maintenance and repair of houses; and enhancing the general cleanliness and hygiene of the kampung (SKIP 1981).

The engagement of the community clearly increased the residents' attachment to the place and their emotional identification, as well as their sense of ownership, which reverted in a more resilient attitude and a stronger community-based organisation for the future actions to develop the area. However, with the termination of the World Bank financing in 1988, it was necessary to continue the upgrading development efforts through a renovated approach and a new program.

In regard to the case of Surabaya, most of the kampung had been upgraded to a certain minimum level due to the KIP projects (Silas 1992). However, because the rest of the city had developed further, better, and faster, the differential between the quality of life in many of the kampung and the affluent sections of the city seemed to have remained unchanged or even worsened. In addition, almost all the previous KIPs mostly focused on improving the physical quality of the kampung environment, ignoring the socio-economic development of these areas (Das 2006). The Comprehensive Kampung Improvement program (CKIP) was launched in 1998 in Surabaya as a joint effort between the City Planning Department (Dinas Tata Kota Daerah, DTKD) and the Laboratory for Housing and Human Settlements (LPP) of the Institute of Technology Sepuluh, which acted as facilitators between the government and the communities. The CKIP aimed, like its predecessors, to improve the quality of housing by comprehensively targeting the area's spatial connectivity and bring about the physical upgrading of the urban environment by providing the services and infrastructure required in the kampung. However, as an innovation, the programme also focused on the community's socio-economic development through community-managed microfinancing and the use of extensive community participation in the planning, implementation and management phases of the project (Swanendri 2002). Based on these objectives, the CKIP proposed five main components to be developed: physical environment improvement; greenery and environmental cleanliness; community development; small and medium-scale business improvement; and, finally, housing improvements.

It is important to note that, like in the previous KIP experiences, the spatial connectivity slum upgrading approach in the new CKIP remained the most important component, without any relevant conceptual modifications and again including streets, footpaths and drainage systems to reduce flooding as the basic elements to improve. However, the funding scheme reflected a new valuable addition to the aim of the programme. Two aspects characterised this new approach in contrast with the first KIPs. The first is that decentralisation led to the local financing of all the projects. The second is that a maximum of 30% of the funds were provided for physical upgrading, while 70% were devoted to developing small business enterprises or housing improvements in the form of microcredits (Das 2006). Due to this new budget distribution, and taking into consideration the importance of the spatial connectivity, we can state that, in this case, the innovative approach was more focused on the process of upgrading the infrastructures than on the upgrading in itself.

In this sense, the CKIP capitalised on decentralisation by transferring decision-making and management responsibilities to new community-based organisations, with the objective of liberating local communities from the tedious administrative structure. Communities were empowered and took on the role of planning, programming, implementing and monitoring the upgrading process with support from the government. Decision-making was done at the community level and through participatory processes based on the identification of their problems, priorities and needs. The programme involved community participation by using the so-called "Three Empowerment" (Tri Daya) sustainability scheme based on physical and environmental development, social improvement, and economic improvement (Septanti 2016). Therefore, we can state that the innovative approach of the CKIP centred on the integrated process, interrelating the physical environment with the social and economic dimensions of the community through the participation of residents towards kampung upgrading. Deriving from this very direct and decentralised empowerment and involvement of the communities in all the steps and activities related to the upgrading of the kampung, a very developed and more sophisticated sense of ownership of the programme was created, which was fundamental to the programme's sustainability, maintenance and management. The premise of formulating programmes and activities according to residents' means and their possibilities for action stimulates the willingness of the community to do more by themselves and work within the limited budgets provided by the city (Dhakal 2002). In this sense, the upgrading of spatial connectivity creates the necessary synergies for the locals to renovate their own houses.

Hence, compared to the previous KIP programmes, it can be stated that the CKIP was more effective in empowering people and communities. Also, the quality of the physical improvements maintained a good standard due to the transmission of the "know-how" after many years of implementation of KIP projects, and also because of the empowerment of the community to decide which physical improvements they needed. However, what limited the effectiveness of the CKIP, as mentioned by Septanti (2004), was, on the one hand, the small budget allocated for the physical improvement components, and on the other hand, the short time given to the community to participate and discuss the projects due to the tight completion schedule linked to the budget, which forced the community to finalise the implementation in a maximum of one year.

7 Concluding Remarks

The urban regeneration of informal settlements should help to integrate them into the formal systems that physically define and politically govern the city. Fifty years after implementing the first KIP, and the later variations incorporated by the CKIP, the Indonesian experience still constitutes a contemporary case of high relevance that fits in perfectly with the current theories and practices of slum upgrading. The KIP case provides lessons for other cities that are aiming to implement an inclusive, fast, simple, replicable and low-cost approach to the resilient urban regeneration of informal settlements.

The significant impact on large segments of the Indonesian population reveals how positive the KIP and the CKIP achievements have been. The spatial connections and infrastructure systems upgrading approach proposed by the KIP suggests a very powerful regeneration tool to increase inclusiveness and to improve the quality of life of kampung residents, bringing to the table some fundamental arguments in the rehabilitation of informal settlements.

- Triggering identity and a sense of pride through spatial connections

The upgrading of the main components of the KIP—streets, footpaths, drain systems, as well as the provision of the basic services needed (education and health facilities)—should be understood as a strategy to generate a broader impact just by activating and planning the public realm of the city. In this sense, the deployment and enhancement of physical connections and social infrastructure not only serves to integrate the informal settlements into the rest of the city through new accessibility and connectivity elements, but it also stimulates the residents to improve their housing on their own. It sparks the reaction to transform all the conditions of the physical environment, motivating more substantial socio-economic transformations.

- The indirect socio-economic benefits of improving a city-wide integrated planning approach

Although land and housing tenure was not one of the primary KIP objectives, the different improvements introduced into the infrastructure of the neighbourhoods have proven to increase the sense of ownership and belonging. In this sense, the introduction of street improvements helps to demarcate the line between public and private land. This delimitation, together with the enhancement of infrastructure, increases the tenure security of individuals and communities, which motivates neighbours to clarify and regularise the status of the land they occupy.

In addition to the improved tenure security, the enhancement of connectivity increases the financial well-being of residents. As a result of the installation of footpaths and roads, there is an improvement in the accessibility to the rest of the city. Thus, enhancement in accessibility and mobility has a three-fold economic effect. It makes access to public transport easier, faster and more efficient, which increases the number of job opportunities and reduces mobility expenses, since travel time is reduced.

Also, mobility enhancement and kampong integration within the city-wide infrastructure has proven to generate social benefits. On the one hand, it results in more inclusivity by providing easier access to markets, schools and health facilities and, on the other hand, it reduces pollution levels in the neighbouring spaces.

- Integrative approach. Blending community, government and financial stakeholders

One of the major reasons for the KIPs success is the integrative approach, which includes a three-fold action program: the combination of bottom-up and top- down approaches; the active participation of the community in the planning and implementation stages; and, finally, the financial integration of several sources including government funds, loans from organisations, and contributions by the community.

It is important to note that an essential aspect of the success of this experience is the implication of community-based organisations in the upgrading process. This commitment creates the stimulating conditions to shift from passive attitudes to active responsibility, encouraging creative attitudes, as well as a positive environment for development.

In this sense, we can state that slum upgrading can only be achieved effectively by linking and integrating the public sector and the community. It has been proven that better outcomes in the physical upgrading can be achieved if the community participates, especially if neighbours are mobilised to contribute with their own resources, which creates socio-economic awareness of the importance of the actions taken.

- Cost recovery is not everything

To finalise, it is important to stress that cost recovery is probably the most relevant handicap of the KIP. As announced by The World Bank (1995), although the program helped significantly to lower urban poverty, it failed in this aspect. Nevertheless, it has to be said that this is a usual weak point for most of the projects dealing with informality and the provision of alternatives for the population living in underserved areas.

From the KIP experience, we learn that, although crucial, the mere implementation of physical upgrading projects and measures does not necessarily promote economic development automatically. The complexity of the upgrading process requires the incorporation of other components that need to be integrated, as we have seen previously.

For this reason, on-site upgrading projects should take cost recovery as a guideline and a framework to follow, rather than a constraint in the upgrading process. Limitations for the sake of cost recovery should be minimised or avoided. We cannot forget the main objective of these projects, placing humans, human needs, and poverty alleviation as the core aim in the pursuit of global progress and equality.

References

Atman R (1975) Kampung improvement in Indonesia. Ekistics 238:216–220

Baross P (1984) Kampung improvement or development?: an appraisal of the low-income settlement upgrading policy in Indonesia. In: Bruno E, Korte A, Mathey K (eds) Development of urban low income neighbourhoods in the third world. Darmstadt, pp 315–333

Budihardjo E (ed) (1992) Sejumlah Masalah Permukiman Kota. Penerbit Alumni, Bandung

Darrundono (1988) The effect of population increase to quality of life: case study Kampung that have been Improved through KIP. Thesis for Postgraduate Programme on Environmental Study—Human Ecology. University of Indonesia, Jakarta

Das AK (2006) What's real and what's rhetorical? The effects of decentralization and participation on slum upgrading in Surabaya. In: Annual conference of the association of collegiate schools of planning, Chicago, 9–12 Nov 2006

Devas N (1981) Indonesia's Kampung improvement program: an evaluation case study. Ekistics 48(286):19–36. Retrieved 2 June 2020

Dhakal S (2002) Surabaya (Indonesia): Comprehensive kampung improvement as model of community participation. Institute for global environmental strategies, urban environmental management project, Kytakyushu

Dhakal S (2003) Surabaya (Indonesia): comprehensive Kampung improvement as model of community participation. Kitakyushu initiative for a clean environment: successful and transferable practices

Djajadiningrat HM (1994) Sustainable urban development in the Kampung improvement programme: a case study of Jakarta—Indonesia. Ph.D. Thesis. Department of Town and Regional Planning, University of Sheffiled

Duncan CR (2004) Civilizing the margins: Southeast Asian government policies for the development of minorities. Cornell University Press, Ithaca and London

Ernawati R, Santosa HR, Setijanti P (2013) Facing urban vulnerability through Kampung development, case study of Kampung in Surabaya, Indonesia. Human Soc Sci 1(1):1–6

Harari M, Wong M (2017) Long-term impacts of slum upgrading: Evidence from the Kampung Improvement Program in Indonesia. NBER Working Paper

Hawken S (2017) The urban village and the megaproject: linking vernacular urban heritage and human rights-based development in the emerging megacities of Southeast Asia. In: Drubach, Lixinski (eds) Heritage culture and rights challenging legal discourses. Hart Publishing

KIP Unit DKI Jakarta (1991) Kampung improvement programme in Jakarta, Bappeda, Jakarta

Kuswardono (1997) Project benefit evaluation: the case of the Kampung improvement program. Master Thesis. Department of Urban Studies and Planning. Massachusetts Institute of Technology

Magis K (2010) Community resilience: an indicator of social sustainability. Soc Nat Resour 23(5):401–416

Mintorogo DS, Arifin LS, Widigdo WK, Juniwati A (2015) Historical old 'kampung' toward sustainable green and clean habitat. In: The international joint conference SENVAR-iNTA-AVAN 2015. Wisdom of the tropics: past, present, Johor, 24 Nov 2015
Municipal Government of Surabaya (2015) Demografi. available at:https://www.surabaya.go.id/berita/8228-demografi Accessed 6 Octover 2019
Poerbo H (1979) The UNEP project in Indonesia, an experience in implementing an integrated approach for slums and marginal settlements. ITB, Bandung
Salim E (1992) Perkampungan ~ ota dan Lingkungan in Sejumlah Masaiah Permukiman Kota, Eko Budihardjo (eds), Penerbit Alumni Bandung, Indonesia. pp 211–220
Sarkawi BH (2015) Improvement of Kampung as an instrument to mitigate floods in Surabaya Humaniora, vol 27, No. 3, Oct 2015, pp. 317–329
Septanti D (2004) Micro-credit system for housing in comprehensive-KIP ad social rehabilitation of slum areas program in Surabaya. J Archit Environ 3(2):1–20
Septanti D (2016) The empowerment of community by C-KIP to improve the slums. Archit Environ 15(1):53–62
Setiobudi A (1990) Kemanfaatan program Perbaikan Kampung BUDP I. Thesis for Undergraduate in Institute Technology of Ban dung (ITB), Bandung
Shirleyana S, Hawken S, Sunindijo RY (2018) City of Kampung: risk and resilience in the urban communities of Surabaya, Indonesia. Int J Build Pathol Adapt. https://doi.org/10.1108/IJBPA-02-2018-0025
Silas J (1983) KIP Program Perbaikan Kampung di Surabaya 1969–1982. Suatu Inventariasi dan Evaluasi. BAPPEM Program Perbaikan PemerintahKotamadya Daerah Tingkat II Surabaya, Surbaya
Silas J (1992) Government-community partnerships in Kampung improvement programmes in Surabaya. Environ Urban 4(2):33–41
Soegijoko S (1985) Managing the delivery of urban services for the poor in Indonesia: case study of KIP in Bandung. Reg Dev Dial 6(2):78–103
Specker M (1981) Low income housing policies and urban development: to the role of the state: a case study of Kampung improvement in indonesia. Thesis for the Degree of Master of Development Studies in Institute of Social Studies—The Hague—Netherlands
Subagio W (1986) Indonesia: country study on urban land development and policies. Haruo Nagamine, urban development policies and programmes, focus on land management. UNCRD, Nagoya, pp 66–111
Surabaya Kampung Improvement Programme (SKIP) Surababya Institute of Technology, Kebalen Community Residents (1981) Kampung Kebalen improvement. Surabaya, Indonesia. Document submitted to the Aga Khan Award of Architecture
Swanendri NM (2002) Evaluation of the Kampung improvement program (KIP) and the comprehensive KIP, Surabaya, Indonesia. Reg Dev Dial 23(1):176–194
Taylor JL (1975/1982) Urban planning practice in developing countries. Pergamon Press, New York
Taylor JL (1987) Evaluation of the Jakarta Kampung improvement programme. In: Skinner RJ, Taylor JL, Wegelin E (eds) shelter upgrading for the urban poor: evaluation of third world experience. Island Publishing Co., Emiel Wegelin, Manila
Turner B (1987) Building community. BCBIHFB, London
UN-Habitat (2006) The world's first upgrading programme. World Urban Forum III. Feature 4/06
World Bank (1983) Learning by doing: world bank lending for urban development. 1972–1982, Washington, D.C.
World Bank (1995) Indonesia impact evaluation report enhancing the quality of life in urban Indonesia: the legacy of Kampung improvement program
Kampung Kebalen Improvement, Surabaya, Indonesia. The Aga Khan Award for Architecture. Retrieved from https://www.akdn.org/architecture/project/kampung-kebalen-improvement
Yudohusodo S (ed) (1991) Rumah untuk Seluruh Rakyat. Inkoppel, Unit Percetakan Bharakerta, Jakarta

Public Space. Urban Regeneration Through Public Space

Various experiences show that public space is a tool to create cohesion and encourage the development of social capital. Urban regeneration through the improvement and provision of public space -as an instrument of social cohesion- promotes a sense of citizenship, generates a sense of belonging to the community that reduces the possibility of conflict, builds the capacity of civil society to organize their own support networks to address gaps, and improves quality of life and environmental conditions, which results on better health of the population.

Regenerating Informal Settlements Through Mapping and Public Space: The Case of BaSECo Compound in Manila

Francesco Rossini

Abstract The city of Manila is experiencing unprecedented urbanisation, which has generated a twin city where the formal and the informal are closely intertwined. Almost 40% of the population in Manila lives in informal settlements. The city has undergone alternate phases of crisis and economic growth, but the urbanisation of its territory has continued uninterrupted. In the near future, the Pasig River Rehabilitation Program will lead to dramatic shifts in the city's fabric, including the relocation of large sectors occupied by informal communities. Under this programme, BaSECo was selected as a priority area for substantial urban renewal. Although this government-led project aims to resettle the residents on-site instead of relocating them outside the city, an integrated approach is needed to implement long- and short term planning strategies and to support decision-making at various stages of the revitalisation process. Using the district of BaSECo as a pilot study, this chapter aims to contribute to the knowledge of community upgrading in Manila by developing a comprehensive plan focused on the improvement of public spaces and community facilities. The aim of this approach aims to incrementally reduce the physical and social segregation of these districts from the broader city and to implement places where social relationships and productive activities can flourish. In particular, this study offers a detailed understanding of three areas of BaSECo compound highlighting the spatial and social characteristics of this dynamic community. This first stage of the analysis is a part of a revitalisation strategy which aims to contribute to the future development of BaSECo by developing a long-term impact via incremental interventions and community empowerment policies. The findings of this research will also be a useful reference for both national and local policymakers working on the revitalization of informal settlements.

Keyword Informal settlements · Open spaces · On-site upgrading · Spatial analysis · Manila · BaSECo Compound

F. Rossini (✉)
School of Architecture, The Chinese University of Hong Kong, Ma Liu Shui, Hong Kong
e-mail: rossini@cuhk.edu.hk

O. Carracedo García-Villalba (ed.), *Resilient Urban Regeneration in Informal Settlements in the Tropics*, Advances in 21st Century Human Settlements,
https://doi.org/10.1007/978-981-13-7307-7_4

1 Introduction

The genesis of the contemporary city has often been characterised by processes of spontaneous evolution that have led to the development of a patchwork of urban fabrics. Some countries, especially in Southeast Asia, are experiencing unprecedented urbanisation, and the existing concepts of urban development may not be sufficient to face the challenges of this unceasing growth. Informal settlements, slums, favelas, shacks, bidonvilles, shanty towns, and squatters are all terms that define some consequences of this unprecedented urbanisation, and which connote negative characteristics and precarious living conditions, such as the lack of basic services and infrastructure, high population density, unhealthy living environments, poverty and high levels of crime (UN-Habitat 2003).

In Metro Manila, the capital of the Philippines, rapid changes in the urban structure have led to an explosion of two opposite, yet related, phenomena: the development of high-density clusters of high-rise buildings; and the formation of dispersed patterns of informal settlements.

These organic and often illegal forms of inhabitation, which are considered parasitic in contrast with the areas that were developed according to a formal planning process, present a surprisingly established order and spontaneous hierarchy, which is not obvious at first sight (Carracedo 2015; Roy 2011).

Large areas in Metro Manila, especially in the downtown area, are occupied by informal communities that will be affected by the city's undergoing redevelopment. The Pasig River Rehabilitation Program (PRRP), a long-term plan to improve water quality and to promote urban renewal, will bring shifts in the spatial pattern of the urban structure, and the informal settlements will have to redefine their role within these important transformations. Although this government-led project is aiming to resettle the residents on-site instead of relocating them outside the city, there is a need to develop a revitalisation strategy that takes into due consideration the interconnected needs of both the local community and the city administration. Undoubtedly, a key aspect for sustainable urban strategy will be the spatial and social integration of these informal areas into the new urban development. As such, a comprehensive analysis examining the spontaneous growth of informal settlements is crucial in order to obtain a clear representation of the urban structure and to gain insight into residents' needs and motivations. Moreover, this understanding of how informal settlements work is key in developing more effective on-site upgrading strategies that are socially, economically and environmentally sustainable.

This chapter focuses on a preliminary study[1] of the informal community of BaSECo, in the port area of Manila, as part of a wider incremental on-site upgrading

[1]This work derives from a project funded by the Social Science Panel, The Chinese University of Hong Kong (Direct Grant for Research), to whom I extend my thanks and acknowledgements. I would also like to acknowledge the students of the School of Architecture of the Chinese University of Hong Kong, who participated in the summer elective course Mapping the Informality, held in Manila in June 2018, for the realization of the drawings and schemes in this chapter. Special mention is due to Cheng Wai Tat, Ma Ting Kwong, Wong Ching Nam Carol, Lau Kai Yin John, Tang Yun Man Kaitlin, Cheung Hiu Yan, Lizhuang Minyi, and Yeung Suk Ting.

plan that uses public spaces and community facilities as the key elements to regenerate informal settlements. Over the last decades, BaSECo has suffered the effects of climate disasters. Because of its location next to Manila Bay and the Pasig River, it is particularly vulnerable to sea and river flooding, as well as storm surges. BaSECo has also endured high levels of crime and poverty, further complicating matters for residents. The study examines three areas of BaSECo that feature different urban conditions and spatial organisations. Urban morphology, typological elements, living conditions and open spaces are analysed using quantitative and qualitative methods in order to gain a comprehensive understanding of the context, identifying tangible and intangible urban phenomena.

One of the most significant challenges to overcome in the regeneration of informal settlements is the lack of data. By analysing the BaSECo compound, this chapter provides a general reflection on the conditions of the informal settlements in Manila in order to propose a methodological and analytical approach that will help implement long and short-term actions and support the decision-making processes at different stages of the revitalisation process.

2 Informality in Metro Manila

Manila is the political and economic capital of the Philippines and one of the largest megacities in the Asia Pacific Region. Despite the fact that city itself accounts for no more than 1.6 million inhabitants, it is actually the epicentre of a metropolitan area that hosts over 12 million people, characterised by extremely high population density. The city was almost completely rebuilt in the second half of the twentieth century, following the devastation of the Second World War, with a very strong impulse from the late 1970s onwards. Over the years, Manila has undergone alternate phases of crisis and economic growth, but the urbanisation of its territory has continued uninterrupted, following no specific spatial order. Rather, urban growth has been marked by highly volatile and chameleon-like configurations (Ortega 2014). This urban explosion does not mean however that there has been a general improvement in citizens' living conditions; quite the contrary, it has accentuated existing inequalities. Despite the fact that the rate of urbanisation is comparable with other countries of the Asia Pacific Region, the city has not experienced the same level of development that usually follows increased urbanisation.

High population growth has put incredible pressure on the basic infrastructure and municipal services, resulting in an overcrowded and polluted metropolitan area (Raflores and Regmi 2015). According to Manasan and Mercado (1999), the deterioration of traffic conditions, the lack of appropriate flood controls and solid waste management, along with interconnected issues of land use, housing, and urban poverty, are four of the most critical challenges facing the city.

A special thanks to Professor Antonio Rinaldi and Valerio Di Pinto who carefully work on the preliminary stage of the configurational analysis.

The need for housing combined with pressure for fast urbanisation hinder long-term and sustainable planning strategies, and over the years this condition has generated a plethora of spontaneous informal settlements. According to Ballestreros (2010), 37% of the population of Manila lives in slums; these settlements are scattered over the entire territory, located along rivers, near garbage dumps, under bridges, alongside industrial establishments and generally wherever there is space and opportunity (UN-Habitat 2003; Ragragio 2003). In the near future, one of the city's major urban transformations will consist of cleaning up the heavily polluted waterways that flow into Manila Bay. This governmental programme is focused on accommodating new flood-mitigation plans (Pornasdoro et al. 2014), which are considered critical for ensuring the city's safety. The plan will cause shifts in terms of the city's spatial pattern, relocating communities away from the city's contaminated canals and rivers (Patiñ 2016).

Previous attempts to rehouse slum dwellers from the riverbanks to more distant locations were not successful, and most of the beneficiaries, finding that they could not make a livelihood on the edge of the city, have since returned to their original locations (UN-Habitat 2003). Relocation policies regarding informal settler families (ISFs) who live in dangerous and high-risk areas will be the upcoming focus of the government's plans. The One Safe Future Program (1SF), one of the housing projects established by the Department of Interior and Local Government (DILG) and involving different stakeholders, offers a people-centred and community-driven approach with the objective of building secure, resilient and sustainable communities. As always in this type of relocation programme, however, moving the inhabitants away from the risk zone is only the first step in a more comprehensive strategy. Relocating them to a new environment, far from their original neighbourhoods, will affect their social relationships and their daily incomes, which were rooted in the informal settlement (Ranque and Quetulio-Navarra 2015). However, not all the informal communities are marked at as high risk; as such, a parallel objective of the government is to integrate such informal areas into current urban regeneration programmes. In this sense, the regeneration of low-income and informal settlements needs to be addressed using alternative methodologies which focus on on-site upgrading schemes that improve the existing conditions of the urban environment and minimise the relocation of local residents.

3 Upgrading Strategies. Public Space as a Method for Urban Regeneration of Informal Settlements

The main limitation hindering the improvement of social and spatial conditions in slums or informal settlements is the lack of genuine political action that focuses on addressing the problems in a comprehensive and sustainable way. In different countries, the bulldozer approach and the relocation programmes of the urban regeneration projects of the 1950s and 1960s did not make substantial improvements due to the complex social and economic problems in informal settlements (Zhu 2009). In the late 1970s, the British architect John Turner (1976) promoted a new approach based on the view that self-help settlements assisted by NGOs had to be perceived as a potential solution for improving slums. Turner's pioneering efforts led to a rethinking of urban regeneration policies in informal settlements, generating various upgrading programmes in different developing countries.

The on-site upgrading approach demands effective engagement and communication with different members of the community. Recent on-site schemes not only focus on improving housing conditions but also on developing open spaces and community facilities as a method of regenerating informal settlements. A participatory process focused on promoting open spaces as social integrators can have 'a strong reliance on the role that public space can play in bringing people together stressing the importance of quality design and architecture' (Riley et al. 2001, p. 527).

As in the case of housing, open spaces in slums are developed by the people (often with the help of NGOs), with the aim of improving the quality of the urban environment. Recent studies have demonstrated that these open spaces play a key role in defining the character of the communities that create and use them (Hernández-García 2013).

According to Jáuregui (2011), planting the 'seeds of urbanism' (in the form of public spaces and community facilities) in the heart of the community may contaminate it in a positive way, improving the physical and social dimensions of the informal district. Physical interventions involving urban space in informal settlements have positive impacts on residents' quality of life by enhancing the general perception of the district (Naceur 2013). In the slum-upgrading context, cases such as those of the Favela Bairro Programme in Rio de Janeiro, the Kampung Improvement Programme (KIP) in Jakarta, or the Community Organisations Development Institute (CODI) programmes in Thailand have become models for transforming slums from illegal settlements into a viable part of the urban fabric (Juliman 2006).

Furthermore, recent researches have highlighted how the use of Geographic Information Systems (GIS) has become a valuable planning tool in informal settlement upgrading (Lefulebe et al. 2015). Assessing and modelling complexity in cities through the quantitative analysis of urban space is gaining momentum, leading to the quest for a new 'science of urbanism'. In this framework, configurational analysis is increasingly engaging scholars and practitioners to rethink how we look at a city, in terms of both its problems and its potential. The first major use of GIS as a planning tool in informal settlements was in the Belo Horizonte project in Brazil (Clodoveu and Fonseca 2011) and in the upgrading project for New Rest and Kanana in Cape Town (Abbott 2001). These pioneering projects have become a model for settlement planning in a number of Latin American countries (Abbott 2000). The use of GIS through the creation of a well-constructed spatial database management system provides an improved and more comprehensive representation of urban settlements, enhances understanding of their evolution over time, and assists in diagnosing and decision-making at different stages of the strategic upgrading process.

4 The Case of BaSECo Compound

The city of Manila is divided into local government units called Barangays, which are the smallest administrative divisions in the Philippines. In the capital alone, there are about 900 of these. BaSECo is part of Barangay 649 and is located at the mouth of the Pasig river, within the Port area of the city (Fig. 1). This engineering island was built on reclaimed land in the mid-1950s under former President Ferdinand Marcos to mitigate flooding along the coastline. For many years, this place was called the National Shipyard and Steel Corporation (NASSCO) and was used mainly as a shipyard.

The name BaSECo, (an acronym for Bataan Shipyard and Engineering Corporation) was established in 1964 when this place was acquired by the Romualdez family. At that time, a few families already lived on this piece of land, but in the 1980s, when BaSECo was formally declared Barangay 649 Zone 68 for administrative reasons, hundreds of families moved in, which also attracted "professional" informal settlers (Mercado 2016). In the past, the area was used as a landfill, especially for ferrous materials derived from port activities. The port still influences this area, creating different job opportunities related to the commerce of waste materials such as paper, glass, plastic, and metals.

In January 2002, the land of BaSECo, under presidential Order No 145, was declared a residential site for the people already living there (GOVPH 2002). Since then, the population has grown steadily from 20,214 to 50,928, making it one of the most populated of the five Barangays in Manila's port area. As explained by Mr. Diocelio B. Candano, a resident of BaSECo and a former member of the Kabalikat, (also known as the People's Organization), after the proclamation many families invited their relatives and acquaintances to live in the new administrative district. This massive population growth has a clear explanation; in fact, once formalized as an official residential site, the former illegal settlement became an area included in the local government's improvement programmes.

Fig. 1 The skyline of BaSECo from the fourth floor of the evacuation centre. *Source* Author

The entire surface area of the district is approximately 52 ha, and, according to the Philippine Statistic Authority (2017), in 2015 the population of BaSECo was approximately 60,000. Over the last 10 years, the area has been affected by a series of fires. In 2002, a blaze left approximately 15,000 residents homeless and then, in 2004, an even more devastating fire occurred, destroying the homes of 25,000 people (DILG 2015). At that time, the government collaborated with two NGOs to provide new homes for the affected population. The residents were engaged in a participatory process, with guidance from Habitat for Humanity and Gawad Kalinga (GK), to build approximately 4000 new houses organised on a regular grid plan (Carracedo 2015).

When the PRRP was established, BaSECo was identified as a priority area for substantial urban renewal. As a result, the community in BaSECo had lived under the threat of eviction for many years. Although this government-led initiative aims to resettle the residents on-site instead of relocating them outside the city, there is still a need to establish a long-term vision for the community. There is considerable attention being paid to the future of this strategic area, and both Kabalikat and Urban Poor Associates (UPA), another local NGO working in BaSECo, are quite active in negotiating with different stakeholders and facilitating the empowerment of local residents. With the involvement of local planners, they drafted the BaSECo development plan with the aim of defining new development areas with minimal reblocking, whilst protecting the families affected in the case of the establishment of environmental preservation areas. These associations were also involved in two projects realised in 2002 and 2010 with the support of the Asian Coalition for Community Action programme (ACCA). This programme was designed to support community-based upgrading initiatives in Metro Manila and, after years of successfully completed projects, it can be said that ACCA has the potential to effectively implement and integrate government policies, especially in terms of co-production and incremental development (Galuszka 2014).

4.1 Methodological Approach to the Urban Regeneration of BaSECo

As mentioned above, this chapter describes the preliminary findings of a more comprehensive academic study which aims to contribute to the improvement of the of BaSECo compound by developing an upgrading plan based on public space as a regeneration strategy. The proposal has three interrelated parts: 'Spatial Analysis', 'Participatory Process', and 'BaSECo Implementation Plan' (BIP). Each part details several key tasks and combines both quantitative and qualitative approaches. This section focuses on the methods used for the survey and spatial analysis conducted in three areas of the district, including the configurational analysis applied at both global and local scales.

4.1.1 Spatial Analysis

Incremental interventions and on-site upgrading approaches rely on a sophisticated understanding of informal settlement forms, as well as their spatial and social structures (Kamalipour 2016). The process of mapping is essential in order to understand the physical context and the sociocultural environment and becomes a vital tool for discussing and planning the revitalisation of the informal community (Carracedo 2016). As part of the spatial analysis, a field survey was conducted by walking through the different sections of BaSECo to obtain an overall picture of the sites under investigation. The digital map based on satellite images that was previously realised was used to establish the survey routes and to record the on- site visits. With the aim of obtaining relevant information on the spatial and social characteristics of each area, the study was focused on the analysis of four aspects:

1) spatial pattern
2) urban form and typological elements
3) the use of the living spaces
4) the analysis of open spaces

Based on previous research, data collection in informal settlements can be very problematic (Barry and Rüther 2005; Chakraborty et al. 2015). Involving residents and NGOs in the data collection process can increase the information gathered regarding the spontaneous growth of these urban configurations. Each site visit was conducted with assistance from the members of Urban Poor Associates (UPA) and Kabalikat. Their support was crucial in the organization of the data collection process, and they provided useful information on the areas studied and assisted the research team in both the mapping analysis and interviews with the community. This approach, known as participatory GIS, allows the community to contribute to the proposed project's body of knowledge through the production and/or use of geographical information (Dunn 2007).

The field research was divided into two interrelated stages. The first consisted of a combination of semi-structured interviews and unstructured discussions with local residents and different stakeholders to understand the community profile. The second step was the direct observation of spatial and social phenomena made through notes, sketching and photo analysis. Datasets such as statistical information regarding population, movement patterns, quality and typology of the urban fabric and open spaces were collected to obtain a comprehensive picture of each settlement and to understand their spatial relationship within the district.

4.1.2 Configurational Analysis

The methods used in this study aim to establish a well-structured knowledge framework and to manage spatial and non-spatial data through the establishment of a comprehensive geographic information system (GIS). Quantitative spatial analysis will be used to increase knowledge reliability and maximise objectivity in understanding the urban structure of BaSECo by stressing the capabilities of configurational analysis: a form-function approach to the city that gives urban space a generative role in social phenomena. Configurational analysis encompasse a wide set of tools and techniques, which share as their fundamental basis the theoretical assumption that form and function in a city are two sides of the same coin: how urban space is arranged and how its single elements are geometrically disposed and mutually connected strongly influence how the space itself is actually used in terms of urban phenomena, such as the distribution of movement flow and activity locations (Hillier 1996). Overall, configurational analysis may be considered a quantitative approach in accounting for how a city operates, as well as in understanding, estimating and governing how people interact in (and with) public spaces.

The spatial data and the configurational analysis indexes, along with other information about the informal settlements (e.g., productive activities and social and religious buildings) and statistical population data, will be integrated in an information and communication technology infrastructure based on GIS (Laurini and Thompson 1996; Burrough and McDonnell 1998). The use of GIS through the creation of a well-constructed spatial database management system will provide an improved and more comprehensive representation of urban settlements, an understanding of their evolution over time, and support for making diagnoses and decisions at different stages of the strategic upgrading process. Different methods of GIS spatial analysis, such as Thiessen polygons, distance matrices, buffer analyses, MCE, and thematic mapping will be applied to provide the framework for the wider upgrading methodology.

5 Mapping Informality, Mapping BaSECo

5.1 Urban Form and Spatial Patterns

The total area of BaSECo covers a surface area of approximately 52 ha; the analysis of the spatial pattern shows a complex urban form where it is possible to identify a critical mass of housing covering more the 80% of the entire area. The main street network, which has enough space to allow for the circulation of people and vehicles, covers approximately 100,000 m^2, which corresponds to 19% of the district. At this stage of the analysis, it is not possible to include minor pedestrian circulation spaces (between 1 and 2 m wide) nested within the intricate network of constructions (Fig. 2). The core is located on the street that formally crosses the district horizontally from east to west. Most of the main public facilities and institutional buildings are located along this street, including the church, the health centre, the evacuation centre, the barangay office, two sports playgrounds, the Benigno Aquino Elementary School, and the President Corazon C. Aquino High School, which stand out for their dimensions in the urban fabric (Fig. 2).

Fig. 2 The main circulation network (left) and community facilities in BaSECo (right). The public schools are clearly visible at the center of the image on the right. *Source* Author

Almost 65% of the spatial pattern in BaSECo has an organic configuration and is located in the southwest part of the district. The areas organised in a more regular pattern are located in the north-eastern part, near the port shoreline and next to the only accessible point of the district (Fig. 3).

These more regular areas were built at two different moments: in 2002, with a community-led housing initiative supervised by the local government (Galuszka 2014); and in 2004, as a result of a reconstruction programme realised under the guidance of two NGOs, Gawad Kalinga and Habitat for Humanity (HfH), after a

Fig. 3 The graphic shows the combination of regular and organic spatial patterns in BaSECo compound. *Source* Author

devastating fire that affected thousands of people, as mentioned above (Gehander and Mörnhed 2008).

5.1.1 Configurational Analysis, BaSECo as a Proximal City

The preliminary results of BaSECo's configurational analysis shows an extremely fragmented urban area with different spatial patterns and urban forms. The urban network of the district presents more than 48,000 nodes per km^2, a value higher than that found in a traditional city. The configurational analysis of such a network unveils a complex structure that is dominated by a dense presence of local centralities, which occupy a large portion of the district and its most densely populated areas (Cutini et al. 2020). Employing the space syntax approach techniques, the analysis of BaSECo compound gives rise to a strong taxonomy of the urban space. At a global scale, the study shows a clear polarization of high values of global integration and choice along just a few lines in the urban network, mostly corresponding to the main roads, where services and community facilities also polarize.This foreground structure (Al Sayed et al. 2014) which supports circulation flows and pedestrian network, revealed that BaSECo doesn't have a consolidated center (Fig. 4). However, the horizontal central spine is perceived by the residents as the most active part of the district with the two public schools, the street market, and one of the biggest playgrounds of the area.

The maps in Fig. 5 show the implementation of space syntax algorithms with different restrictions to calculation intervals known as local radii (Hillier 1996; Turner 2001). The local analysis at 200 m radius (Fig. 5, left) revealed the presence of two highly integrated urban areas, located in the south and northeast of the settlement. The granularity of the local configurational pattern is not clear in the metric analysis due

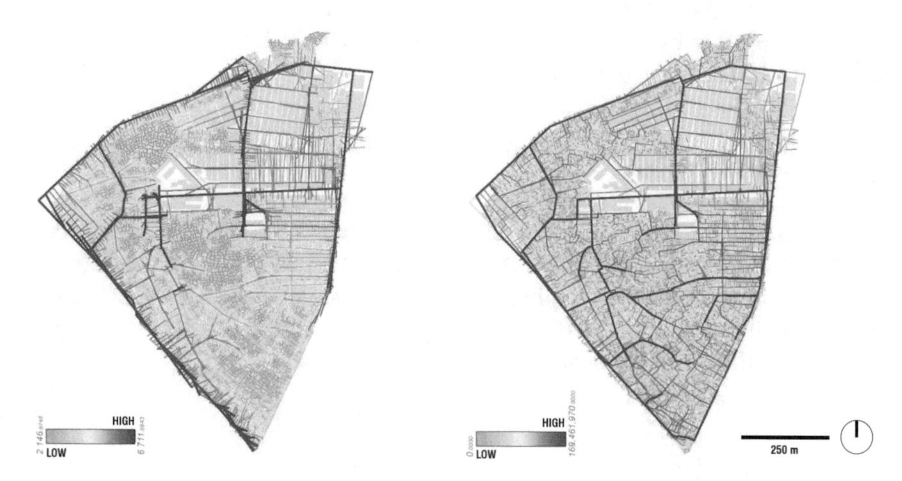

Fig. 4 The global configuration of BaSECo: (left) Global Integration-fat, red lines highlight the so-called 'integration core'; (right) Global choice. Light, grey, solid hatches outline closed spaces. *Source* Cutini et al. 2020

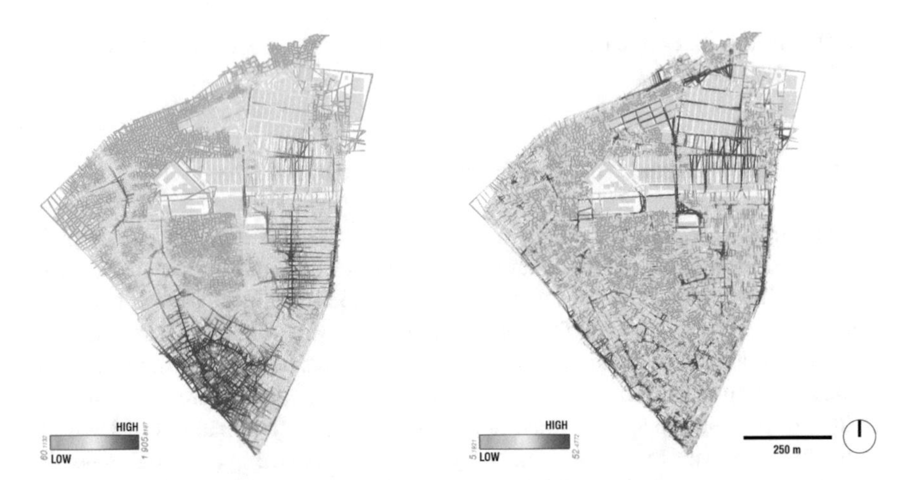

Fig. 5 The local configuration of BaSECo: (left) Local Integration-metric radius 200; (right) local integration-topological radius 3 steps. Light, grey, solid hatches outline closed spaces. *Source* Cutini et al. 2020

to the excessive gradient between integrated (fat red and orange lines) and segregated (green and blue lines) nodes of the urban graph. However, the topological analysis (Fig. 5, right) clarified the results of the local analysis indicating, especially in the south part of the district, a high number of well-separated local centralities (Cutini et al. 2020). The combination of both global and local analysis shows that a large part of the urban structure of BaSECO compound is constituted by an archipelago of 'small villages' linked by an intricate pedestrian network, where the 'proximity' of people and the density of social activities are the key elements of the district. In this regard, the definition of BaSECo as a 'proximal city' aims to takes a step forward in theory and practice by facilitating the elaboration of alternative upgrading strategies to address the challenge of urban informality.

5.1.2 Spatial Analysis

As mentioned earlier, the study is based on the analysis of three specific areas within the BaSECo compound: Dubai Site (DS), New Site (NS), and Gawad Kalinga (GK). These areas present different spatial patterns: DS delineates a more spontaneous and organic configuration, while NS and GK are based on a more regular grid, despite substantial differences between them (Fig. 6).

5.2 Gawad Kalinga

- Spatial Pattern

Fig. 6 Highlighted in black the districts of Gawad Kalinga (left) Dubai Site (centre) and New Site (right). *Source* Author

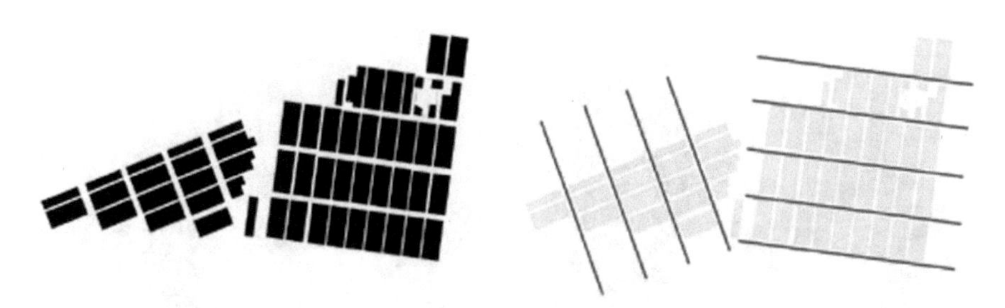

Fig. 7 The layout of GK based on a regular grid pattern. *Source* Author

The area of GK was completely rebuilt after a fire in 2004. The district is characterised by an orthogonal layout marked by regular rows of houses and narrow streets. GK was developed over an area of about 4 ha and its name derives from the NGO Gawad Kalinga that was involved in the housing reconstruction, providing affordable solutions for thousands of people affected by the fire. The analysis revealed that this grid pattern allocates almost 16% of the area to open space,[2] representing an average of about 1.3 m^2 of outdoor space per person. Due to the space constraints of the site, the rows of houses are aligned in two different directions to provide more living units (Figs. 7 and 8).

External circulation occurs on a linear grid of 5-m streets, while, internally, the rows are organised on a grid of 1.6-m passageways that provide access to each unit. The houses are separated by a back alley of less than 1 m, which is used for various activities such as laundering, cooking, and washing, representing an extension of the interior living space. The community is organised into 'villages' formed by four rows of houses; walking through the alleys of GK district it is possible to identify the names of the sponsors—or, better yet, partners—who contributed to the construction of entire rows or villages.

- Urban Form, Materials and Typological Elements

[2]GK area 40,991 (plot 27,127 + 30% street 6000).

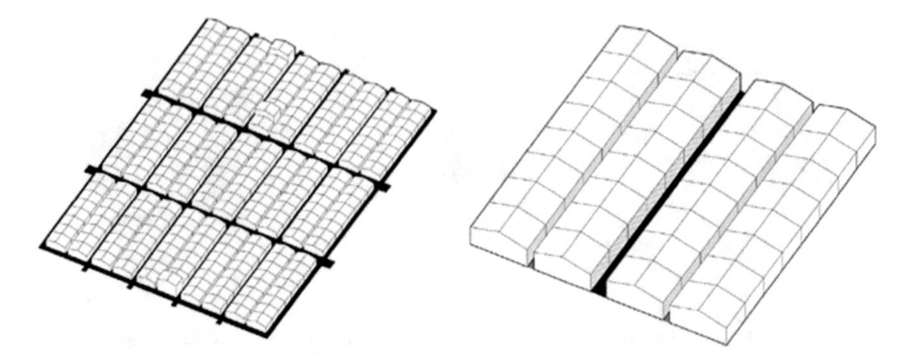

Fig. 8 The organisation of GK district based on row houses which consist of singular units of 20–24 m^2. *Source* Author

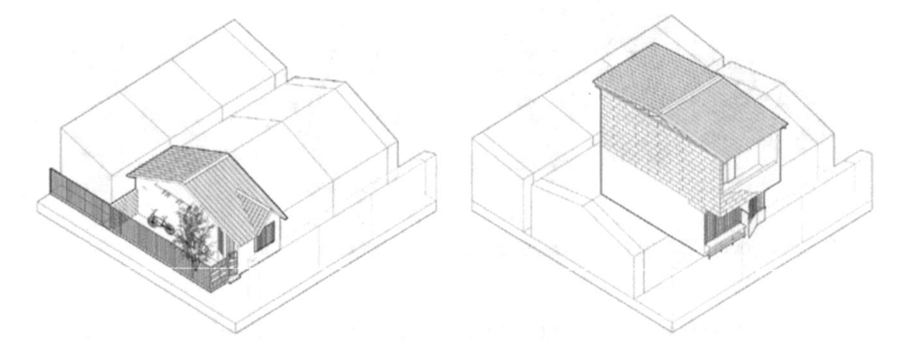

Fig. 9 Two typologies of housing in GK: the yard unit (left) and the two-floor unit (right). *Source* Author

The rows of houses, constituted by eight units, were originally built one storey high, but according to the original plan, residents may upgrade their houses by adding one or two storeys.

Based on the site analysis, three main typologies of houses were identified: the regular unit, the two-floor unit; and the yard unit (Fig. 9). The first two are quite similar and, at the time of the site visit, only a handful of units had an additional floor, which often cantilevered 1 m, creating an intermediate covered area above the entrance. This additional floor, like the one-unit house itself, is usually made using external partitions in concrete blocks with a roof made from a wood structure and corrugated steel sheet. The yard unit is usually located in the corner at the end of the row and is made up of an outdoor area running lengthwise alongside the unit. This yard is protected by a wooden or steel fence and provides a semi-private open space commonly used as a garden, storage or laundry area.

The personalisation of the living environment is a recurring feature in informal settlements; however, according to Gehander and Mörnhed (2008), the NGO Gawad Kalinga encourages inhabitants not to make exterior alterations to the houses, although they can customise their living spaces by adding partition walls, floor finishings and a ceiling. One of the features that makes any Gawad Kalinga village clearly recognisable, including the one in BaSECo, is the use of bright colours used to paint the external walls of the houses. Green, blue, yellow, and pink/red were the most common colours found during the survey.

- Use of the Living Space

In GK district, the standard dimensions of each unit are 5 m by 4 m, which generates a living space of about 20 m^2 per family. According to the survey, family units averaged about five members. As mentioned previously, only a few units are two stories high, and usually this typology includes the living room and kitchen on the ground floor with the bedrooms on the upper floor. In many cases, the inhabitants expanded their units by extending the roof into the common alley. This covered outdoor space creates an intermediate domain used for gardening, child recreation, resting or simply as a space to dry clothes. This appropriation of the common space is perhaps the norm in informal settlements; however, it is also the most peculiar aspect in that it influences how communal areas are perceived. Even though these circulation spaces are narrow and cramped, the spatial quality is above standard compared with other areas in BaSECo, in that almost all the alleys are paved and during the frequent flooding these passageways seem relatively protected. Actually, in the reconstruction plan for this area, donations were used not only to build housing units but also to pave streets and plant trees with the aim of creating overall improvements in the living environment (Gehander and Mörnhed 2008).

- Analysis of Open Spaces

Social life often takes place in the alleys, an intimate space that is in direct contact with the interior of the houses and which becomes the place where the community socialises (Fig. 10). These alleyways represent the collective space of GK, encouraging the formation of micro-communities within the district. These linear open

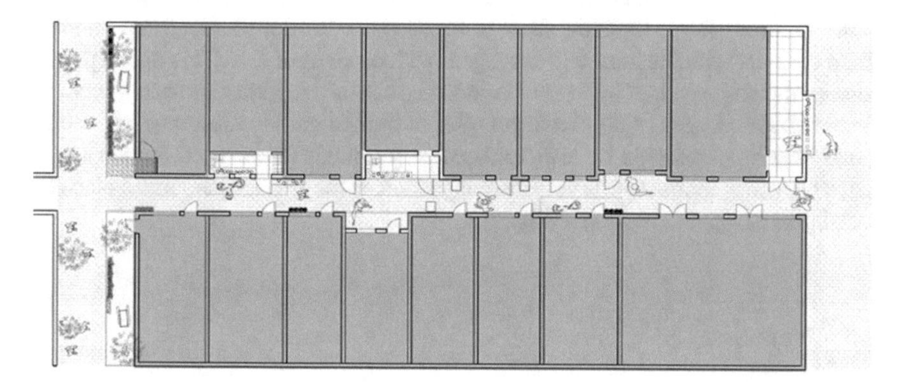

Fig. 10 The image shows how in GK district the back alleys are used as an extension of the living space. *Source* Author

Fig. 11 One of the biggest open spaces in GK district. As the image shows, this open space is not well maintained. *Source* Author

spaces are animated by the sari-sari, the convenience store in the Filipino community where it is possible to buy candy, cigarettes and canned goods. Often, the front of the house is occupied by these typical local stores, which comprises a common form of income for many families, while providing basic daily necessities to other residents in the alley.

Located in the northern part of the district, the multi-purpose hall is the heart of the GK community. This facility, which has a large courtyard that has recently been equipped with a canopy, is administered by the senior members of GK. The hall hosts various social activities for the community and for local children, who can play in a larger but controlled space. Just beside the multi-purpose hall there is perhaps one of the widest open spaces in the area. This place is not well maintained and stands in sharp contrast with the carefully managed outdoor areas of GK. This open space is colonised along its edges by various activities, such as temporary sari-sari stores and street vendors. Cages of chickens and piles of burned garbage surround the central space which is occupied by a dilapidated playground. However, this unmanaged space has the potential to alleviate the congestion of the area and become a place for social gatherings and meetings (Fig. 11).

5.3 New Site

- Spatial Pattern

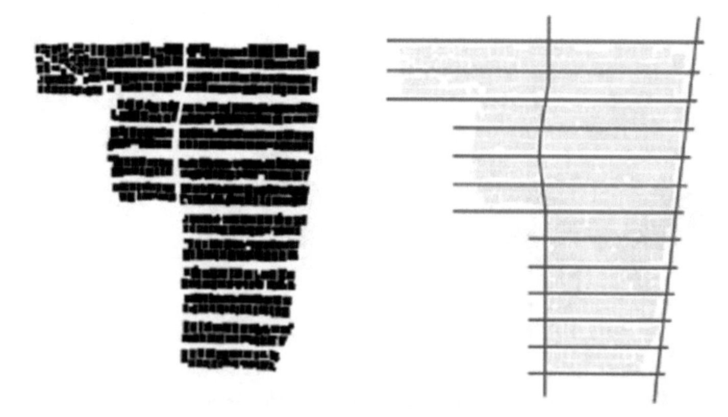

Fig. 12 The regular grid pattern and circulation network in NS. *Source* Author

The district of NS occupies 4.8 ha of land and is located in the east part of BaSECo, near the shoreline. In 2002, with help from Technical Assistance and Organization (TAO Philippines), the community took part in participatory planning sessions to develop the layout of this area. The district, which also includes the Barangay hall, was developed under the supervision of the local government, which donated materials worth up to PhP 15,000 per family for the construction of 205 houses, built with the help of specialists (Galuszka 2014). The layout of NS consists of a grid plan based on a regular urban form; 12 groups of linear blocks of one, two and three levels are intersected by 13 streets that are 5 to 6 m wide (Fig. 12). These streets cross the area from west to east allowing for the circulation of vehicles and people.

The district is traversed from south to north by two main roads that connect NS to other key areas of BaSECo. Looking at the figure ground, this grid layout allocates almost 28% of the area to open space and streets,[3] representing an average of about 2.4 m^2 of outdoor space per person.

- Urban Form and Typological Elements

In NSthe most common building typology is the two-floor house made of concrete and an external wall in hollow blocks. This typology presents different modes of customisation in terms of decorations or construction elements. Openings and balconies do not have standard dimensions, colours or finishing materials. The staircase can be placed internally, serving one family, or externally if the building belongs to two different families who live on separate floors. Some of the tallest constructions in BaSECo compound are located in NS; these consist of three- and four-storey houses which belong to the wealthiest residents (Fig. 13). The façades of these buildings, similar to those of the two-floor homes, have different sorts of decorations with balconies and sometimes even a terrace on the rooftop. The balconies are usually

[3]NS area 48,113 (plot 35,063 street 13,050).

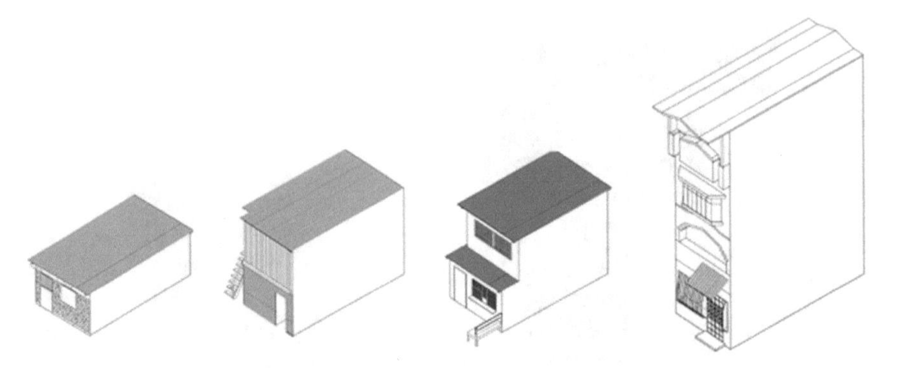

Fig. 13 The image shows the different typologies of housing in NS district. They vary in height from one to fourth-story buildings. *Source* Author

integrated within the façade, without a cantilever over the street, and are normally used to hang clothes as well as to provide an extension of the living space.

It has to be noted that not all the buildings in this area are in good condition; the poorest families live in one-floor units built out of concrete blocks, recycled wood and corrugated steel sheets. Due to the fact that they only cover the basic necessities, people living in these basic living spaces generally prefer to spend their time outdoors, extending their daily activities into the common area of the street.

- Use of the Living Space

On average, the living space in NS is relatively generous: the buildings are developed on linear plots that vary from 3.5 to 4 × 6 to 9 m, which generate spaces that range from 20 to 36 m^2 per floor. It was observed that some buildings occupy two plots, which increases the possibility of having other businesses associated with the housing space. The bedrooms are usually on the upper floors, while the ground floor typically includes an open kitchen and a living room, which often extends outdoors with a sari-sari store. These local shops are normally run by the women of the household, as they are able to take care of the shop and the children without leaving the house. This storefront also serves as a filter between private and public space, creating hubs of socialisation along the streets.

- Analysis of Open Spaces

The prominent location and the proximity to the Barangay hall, the schools, and the coastline make this district more permeable to traffic flows. Most of the activities, such as barber shops, small restaurants, hardware stores, and electronic shops are located on the main street that flanks NS on the north side. The dimensions of streets within the district vary from 5 to 8 m, facilitating the creation of intermediate spaces on either side, which can become the setting for local residents' daily activities. Often, the streets in NS become spontaneous basketball courts where the teenagers play around an improvised hoop. The sari-sari shops along the streets become, as

Fig. 14 The image shows the typical street life in NS district. The presence of trees helps to cool down the temperature during the hottest hours of the day. *Source* Author

described, not only a place where daily necessities can be purchased, but also a place to chat with neighbours, to have a quick lunch or to quench one's thirst with a drink. Street food vendors were also observed lingering on the street corners, contributing even further to the liveliness of this area of BaSECo.

In NS, despite relatively good housing conditions, not all circulation spaces are levelled and paved and, in some cases, it is quite difficult to walk in this area due to the precarious state of the streets. Vegetation is rarely present in this area, but the few trees identified can provide sufficient shade to significantly lower the temperature, thus contributing to a more comfortable climate (Fig. 14). Unlike GK, NS does not have a proper open space where the community can meet and linger. However, in NS there is the office of Kabalikat: a two-storey building located in close proximity to the Barangay office. This association, with over 1500 members spread across various areas of BaSECo, offers different community services for its affiliates. Kabalikat owns a learning centre in the nearby Dubai Site, where the community can organise various social activities and which often serves as a meeting place for the district's decision-making.

5.4 Dubai Site

- Spatial Pattern

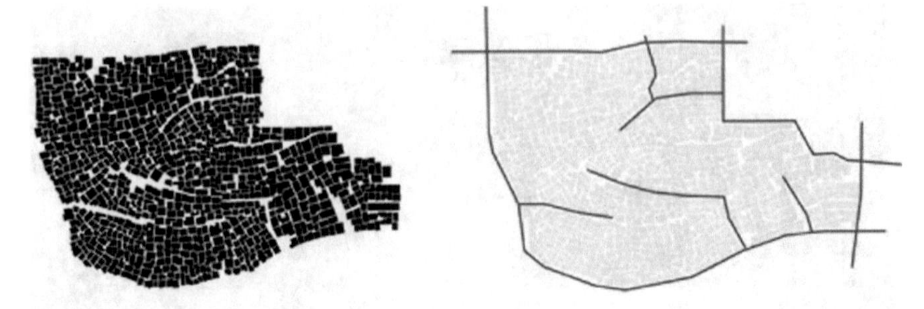

Fig. 15 The organic spatial pattern of DS (left) and the main street network (right). *Source* Author

According to the spatial analysis, DS is perhaps the poorest district when compared to the NS and GK. This spontaneous and organic area is located at the core of BaSECo, at the southern part of the district's main facilities (Fig. 15). DS occupies an area of 6.8 ha, but this intricate settlement allocates only 8% of the area to circulation space,[4] which represents an average of only about 0.6 m^2 of outdoor space per person. This district has not been structured by any plan; however, the local government is providing new homes organised on a linear plan for 225 families, as part of the public safety and relocation programme, which intends to reduce disaster vulnerability and displacement of poor, urban communities (DILG 2015).

Given its spontaneous and organic form, it is very difficult to establish orientation in this area. The maps obtained from the digitalisation of satellite images have certainly facilitated the spatial analysis of this area, which is continuously being modified by changes and alterations.

- Urban Form and Typological Elements

Most of the houses in DS are made up of mixed structures, one or two stories high and characterised by the use of multiple materials, generally recycled from the waste of various sources. These typical informal constructions, which constitute almost 65% of the organic pattern in BaSECo compound, comprise different spatial solutions that serve to accommodate the basic needs of the inhabitants (Fig. 16). Often, the two-storey houses are shared by two families having separate access points. The ground floor is made up of concrete blocks or wood, while the upper floor is usually built only in timber in order to prevent excessive weight and overheating problems. If not recycled from a dumping area, the typical wooden material is coconut lumber, which is cheap, durable and allows for flexibility in construction.

External partitions are layered with different light materials such as metal and plastic sheets, bamboo and wooden plates, canvas, and welded steel sticks. The roof structure is made of timber and finished with corrugated metal sheets, sometimes with the addition of reflective sheets and plastic canvas to prevent heat and leakage inside the living space. Different materials are connected by mechanical fasteners like nails and cement, or even simply tied with rope or rubber. Openings may vary

[4]DS area 60,360 (plot 56,230 (43,254 + 30%) streets 4130).

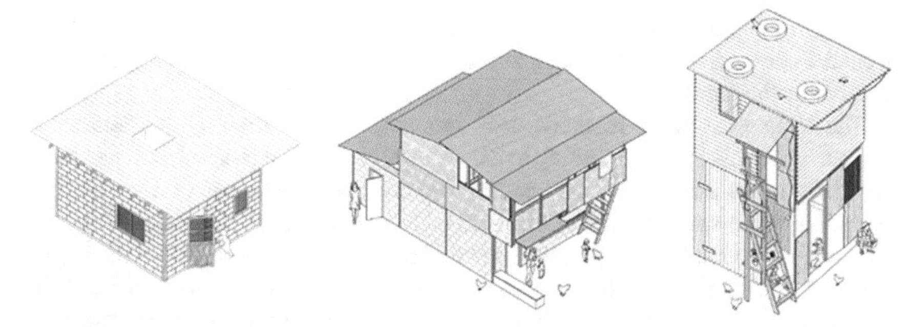

Fig. 16 Different typologies of houses in DS district. *Source* Author

in their dimensions and are usually oriented to make the best use of the daylight and according to the family's needs. In terms of privacy, crafted concrete blocks are normally used to provide ventilation for the toilets.

- Use of the Living Space

Due to the lack of planning, this district presents various spontaneous spatial solutions; for example, the shape of the houses is not regular but rather generated according to the site's spatial constraints. The living spaces in this area of BaSECo are comparatively smaller, with an average of 6–10 m^2 per floor, with the height of the ceiling ranging from 2 to 2.3 m (Fig. 18).

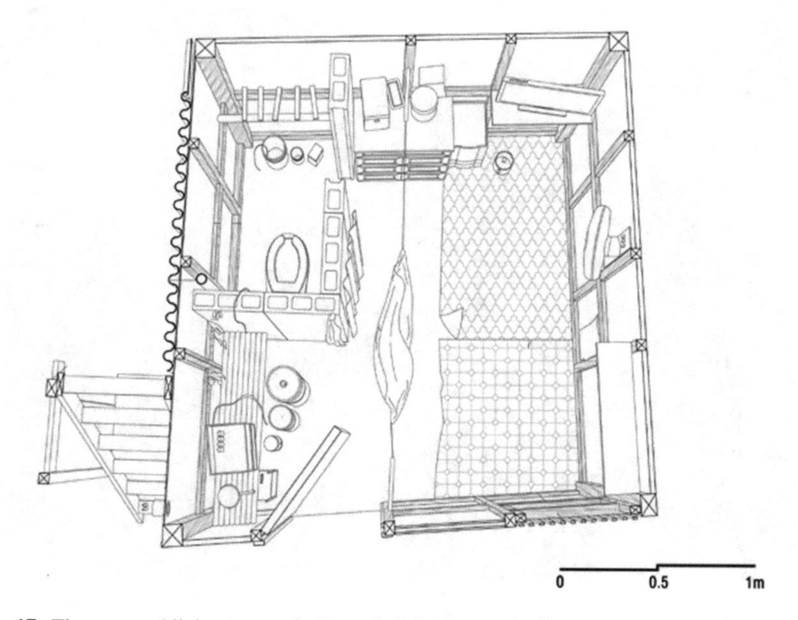

Fig. 17 The cramped living space of a house in DS. *Source* Author

Similar to what was observed in GK and NS, in some cases a small portion of the ground floor is used to sell snacks and drinks as a means of generating additional income for the family.

The simple assembly methods used for these constructions allow families to upgrade the living space according to the needs of the inhabitants. Internal partitions are easily made using fabric and ropes. The original structure is easily expanded by adding more space to the living room or to make a new bedroom for the kids. This organic, flexible, and adaptive form of urban growth represents a distinctive aspect of this district, which in fact can also be found in different areas of BaSECo. Unfortunately, the hygienic conditions of DS are not at all satisfactory: there are often flooding problems due to difficulties in the disposal of blackwater. In fact, it is quite common in DS for floors to be raised in order to protect the living space from this issue.

- Analysis of Open Spaces

Most of the alleys and narrow streets in DS are unpaved, making circulation in this area quite problematic. However, the survey showed that public life flourishes in corner spaces left unoccupied by constructions, where pool tables, video games, karaoke stations and improvised basketball courts animate scattered places within the district. These points of activity, spontaneously created outside housing units under cantilevers or temporary canopies, constitute an unstructured form of social aggregation at different hours of the day (Fig. 18).

Fig. 18 A common social activity in DS. *Source* Author

To escape from the congestion of this area, teenagers and kids flow into one of the biggest outdoor spaces, located in the north of DS. This public playground is meant to serve the whole community of BaSECo, but, due to its proximity to DS, most of the local people consider this place as an extension of their district. This partially covered space, which contains a playground and a basketball court, is usually overcrowded around lunchtime and in the late afternoon when children are back from school. Another observation of the area is that of the plentiful activity along the edge of the playground and around the schools, such as street vendors and informal market activities scattered up and down the streets at different times of day.

6 Considerations and Discussion

Making cities and human settlements safer and more inclusive, resilient, and sustainable is at the core of Goal n.11 of the UN-Habitat. Slums and low-income settlements are often associated with negative characteristics and precarious living conditions. However, there are also significant positive aspects to informal communities that can serve as sustainable solutions to the uncontrolled urban growth in emerging countries.

As previously described, BaSECo district is affected by structural weaknesses that must be addressed with a comprehensive upgrading strategy. The need for housing, as well as the extremely high population density, have put incredible pressure on basic infrastructure and community services. At first sight, BaSECo presents all the typical characteristics of any informal community such as congestion, poor hygiene conditions, misaligned street spaces and a lack of urbanity. However, the three districts analysed present substantial differences in terms of urban form, social activities, use of living spaces, and open spaces. The study of GK, NS and DS revealed that 'informality' takes on distinctive forms according to varying social and spatial conditions in the urban environment. The analysis of DS is representative of the logic that defines the organic spatial patterns overall but is not exhaustive, as it is necessary to take into consideration other parameters. It may be possible to find the same construction typologies and the same use of open spaces, but the social dynamics can vary depending on the specific surroundings of the local context.

In DS, the hygienic conditions found were substantially worse than in GK and NS, with unpaved circulation spaces, flooding problems, and rubbish littering the streets in many areas of the district (Fig. 19). The lack of structured planning in this settlement has resulted in a dense urban area with cramped living spaces and a low index of open spaces per person (0.6 m^2). This parameter is extremely low, even compared to GK and NS, but despite the overcrowded environment, the outdoor spaces reveal grassroots social activities that are deeply integrated within the urban fabric.

As described, GK and NS are structured on a regular layout; however, this spatial arrangement comprises two very different urban settlements. In GK, the grid carves out a system of passageways that have been converted into an extension of the interior living space by local residents. The close proximity of the linear rows of houses and

Fig. 19 The problematic conditions of the circulation spaces in DS. *Source* Author

the narrow dimensions of the lanes results in a sort of gated community, where these circulation spaces tend to have an intimate character. However, this regular layout, which provides a good standard of living, also serves to check the spontaneous occupations typical of informal settlements, which reduce the standards of public space and obstruct its purpose.

Unlike that of GK, the NS grid is very permeable to urban flows of both traffic and people and comprises a settlement based on the street as the 'urban rule' used to align the constructions. The width of the streets subsequently allowed for the development of taller buildings, resulting in some of the safest constructions built in BaSECo, thanks to the support of the local government. As mentioned previously, some of the units are even three or four storeys high and equipped with toilets (Galuszka 2014). This planned layout forms a unitary but flexible organism that leaves some degree of freedom in terms of the construction of different housing typologies. The streets, dotted with small commercial activities and places to cultivate social relationships, are the principal open spaces of this district (Fig. 20). The different spatial patterns and urban forms also influence societal relationships and the way in which open spaces are used. In GK and NS, a regular layout has fostered the development of linear communities, in that this type of organisation facilitates social interactions in the streets, thereby consolidating a sense of belonging among residents. By contrast, in the organic pattern of DS, neighbourhood relations are not so clearly defined, and social interaction occurs among small groups of families essentially because of proximity.

In all the areas analysed, the lack of open space and community facilities is a serious issue to be addressed. In most cases, and specially in DS, the unplanned and irregular structure, typical of informal settlements, generates dynamic spaces where distinctions between public and private are not clearly defined. The proximity of people and open space in informal settlements generates a specific relationship,

Fig. 20 The streetscape in NS where the sari-sari stores create places of social activities. *Source* Author

'because they are socially produced and constructed' (Hernández-García 2013, p. 5). Such intermediate open spaces are not simply vital to the effective functioning of settlements, they offer a place where social relationships and productive activities can flourish, providing what is perhaps the greatest value to informal communities (Fig. 21).

As argued by Carracedo (2016), the creation of new open spaces through the active participation of the community can help to enhance the social integration of neighbors, as well as achieving a general improvement of the urban environment. However, the approach to this strategy might differ depending on the morphology of the urban fabric. In fact, as demonstrated by the spatial analysis, the spontaneous and porous urban tissues of DS allow for a strategy of selective on-site relocation and densification, realigning the new constructions to create also open spaces. Differently, in both GK and NS the grid system and the alignment of the buildings do not provide enough flexibility to incorporate new open spaces without affecting the general urban structure.

Fig. 21 The spontaneous use of the open spaces in DS. *Source* Author

7 Concluding Remarks

Manila is experiencing a globalised form of gentrification. This recent urban development model focuses on the partnerships between the public and the private sector, to realise the vision of a globally competitive and modern metropolis (Ortega 2016). Over the years, the national government has developed a number of policies and programmes in response to the challenges of urban informality and to contain the expansion of new slums in the territory, but these have generally failed as sustainable, long-term plans. However, successful revitalisation plans in Brazil, India, Thailand, Indonesia and South Africa have demonstrated that upgrading strategies are a viable, low-cost and effective way to help the urban poor solve problems in their communities (UN-Habitat 2009). The participation of the different stakeholders at every stage of the upgrading process is an essential task when working in the complex environments of informal settlements because it invites people to respond creatively to a given problem and develop solutions that better meet local needs. Furthermore, these examples not only propose new housing; they also focus on the development of streets, public spaces and other facilities as a strategic approach to urban revitalisation (Carracedo 2015). The study presented in this chapter unfolded the unique DNA of BaSECo compound. The analysis across different scales revealed sophisticated urban phenomena as well as hidden and diverse socio-spatial dynamics. Besides, it demonstrated the need for a comprehensive understanding of these complex urban settlements to acknowledge the social, economic, and cultural value of the community. The aim of this research is contribute to the knowledge of on-site upgrading approaches by formulating a sustainable and replicable urban process based on the incremental implementation of public spaces and community facilities. The findings will also help address the tension between bottom-up and top-down approaches by providing government departments with useful recommendations for the revitalisation of informal settlements.

References

Abbott J (2000) An integrated spatial information framework for informal settlement upgrading. In: International archives of photogrammetry and remote sensing, vol XXXIII, Part B2. Amsterdam

Abbott J (2001) The use of spatial data to support the integration of informal settlements into the formal city. Int J Appl Earth Obs Geoinf 3(3):267–277

Al Sayed K, Turner A, Hillier B, Iida S, Penn A (2014) Space Syntax Methodology, 4th edn. Bartlett School of Architecture, UCL, London

Ballestreros MM (2010) Linking poverty and the environment: evidence from slums in Philippine cities. Discussion Paper Series No. 2010-33, Philippine Institute for Development Studies

Barry M, Rüther H (2005) Data collection techniques for informal settlement upgrades in Cape Town. South Africa. URISA Journal 17(1):43–52

Burrough P, McDonnell R (1998) Principles of geographical information systems. Oxford University Press, Oxford

Carracedo O (2015) Spatial and organization patterns in informal settlements. In: A morpho-typological approach. 22nd ISUF conference: city as organism. New visions for urban life, Rome

Carracedo O (2016) Spatial and organization patterns in informal settlements. In: Amorphotypological approach. 22nd ISUF conference: city as organism. New visions for urban life, Rome. ORO: Singapore

Chakraborty A, Wilson B, Sarraf S, Jana A (2015) Open data for informal settlements: Toward a user's guide for urban managers and planners. J Urban Manag 4(2):74–91

Cutini V, Di Pinto V, Rinaldi AM, Rossini F (2020) Proximal cities: does walkability drive informal settlements? Sustainability 12(3):756

Clodoveu A, Davis Jr, Fonseca F (2011) National data spatial infrastructure: the case of the Brazil. InfoDev/World Bank, Washington D.C. Retrieved from: http://www.infodev.org/ publications

Department of the Interior and Local Government (DILG) (2015) BaSECo compound families to receive new home. Official Gazette GOVPH. Retrieved from: http://www.officialgazette.gov.ph/2015/10/30/BaSECo-compound-new-building/

Dunn CE (2007) Participatory GIS — a people's GIS? Progress Human Geography 31(5):616–637. https://doi.org/10.1177/0309132507081493

Freeman L (1978) Centrality in social networks conceptual clarification in Social. Networks 1(3):215–239. https://doi.org/10.1016/0378-8733(78)90021-7

Galuszka J (2014) Community-based approaches to settlement upgrading as manifested through the big ACCA projects in Metro Manila, Philippines. Environ Urban 26(1):276–296

Gehander M, Mörnhed E (2008) From slum to adequate homes: a study on housing solutions for the urban poor in Manila, Philippines. Lund University, Helsingborg. Unpublished Thesis

Government of The Philippines (GOVPH) (2002) Proclamation No. 145, s. 2002|Official Gazzette of The Republic of The Philippines. Retrieved from: http://www.officialgazette.gov.ph/2002/01/18/proclamation-no-145-s-2002/

Hernández-García J (2013) Public space in informal settlements: the barrios of Bogotá. Cambridge Scholars Publishing

Hillier B (1996) Space is the machine. UK: Cambridge University Press.

Jáuregui JM (2011) A project for the favelas of Rio de Janeiro. Archphoto

Juliman D (2006) The world's first slum upgrading programme. World Urban Forum III: An International UN-Habitat Event on Urban Sustainability, Vancouver
Kamalipour H (2016) Forms of informality and adaptions in informal settlements. Archnet-IJAR Int J Archit Res 10(3):60–75
Laurini R, Thompson D (1996) Fundamentals of spatial information systems. Academic Press, New York
Lefulebe B, Musungu K, Motala S (2015) Exploring the potential for geographical knowledge systems in upgrading informal settlements in Cape Town. South Afr J Geomat 4(3):285–298
Manasan RG, Mercado RG (1999) Governance and urban development: case study of Metro Manila. In: PIDS Discussion Paper Series No. 99-03, Philippine Institute for Development Studies
Mercado RM (2016) People's risk perceptions and responses to climate change and natural disasters in BASECO Compound, Manila, Philippines. Procedia Environ Sci 34:490–505
Naceur F (2013) Impact of urban upgrading on perceptions of safety in informal settlements: case study of Bouakal, Batna. Front Archit Res 2:400–408
Ortega AAC (2014) Mapping Manila's mega-urban region. Asian Popul Stud 10(2):208–235
Ortega AAC (2016) Manila's metropolitan landscape of gentrification: global urban development, accumulation by dispossession and neoliberal warfare against informality. Geoforum 70:35–50
Patiñ PI (2016) Building resilient and safe communities against poverty and disaster. Asian Cities Climate Resilience, IIED
Pornasdoro KP, Silva LC, Munárriz MLT, Estepa BA, Capaque CA (2014) Flood risk of metro manila barangays: a GIS based risk assessment using multi-criteria techniques. J. Urban Reg Plann 2014:51–72
Raflores L, Regmi R (2015) Understanding the water and urban environment of a megacity: the case of Metro Manila, Philippines. In: Water and urban initiative. Working paper series. United Nations University
Ragragio JM (2003) The case of Metro-Manila. In: Understanding slums: case studies for the global report on human settlement 2003: UN-Habitat/Development Planning Unit (DPU), University College London
Ranque L, Quetulio-Navarra M (2015) One safe future in the Philippines. In: Forced migration review. Disasters and displacement in a changing climate, pp 50–52
Riley E, Fiori J, Ramirez R (2001) Favela Bairro and a new generation of housing programmes for the urban poor. Geoforum 32(4):521–531
Roy A (2011) Slumdog cities: rethinking subaltern urbanism. Int J Urban Reg Res 35(2):223–238
Sabidussi G (1966) The centrality index of a graph. Psychometrika 31(4):581–603. https://doi.org/10.1007/bf02289527
Turner J (1976) Housing by people: towards autonomy. In: Building environments. Marion Boyars, London
Turner A (2001) Angular analysis. In: (Proceedings) Proceedings 3rd International Symposium on Space Syntax. Georgia Institute of Technology, GA, pp 30.1–30.11
UN-Habitat (2003) The challenge of slum: global report on human settlements 2003. London and Sterling, VA
UN-Habitat (2009) Slum upgrading facility: exchange visit to the community organisations development institute in Thailand. Nairobi
Zhu H (2009) The research of upgrading models of informal cities in developing countries. World Acad Sci Eng Technol 30:1565–1568

Medellín, Urban Renewal of Informal Settlements Through Public Space: The Case of the North-Eastern Integral Urban Project (PUI)

Armando Arteaga

Abstract This chapter explains and highlights one of the most important urban renewal tools used in the city of Medellín to tackle informality: the Proyecto Urbano Integral (PUI) [Integral Urban Project]. Looking at the experience of the first PUI, tracing its origin, and considering the practices and experiments developed over three decades to address informal urban settlements, this article analyses the impact of the PUI as a planning tool and how the use of public space in Medellín became the foundation for one of the main urban upgrading strategies. A decade after their implementation, the article investigates the strategic role of PUIs in establishing a method for urban renewal in informal areas, as well as their value as pilot and replicable experiences throughout the city. The investigation analyses 23 public space projects to evaluate their performance and their real social and physical impact.

Keywords Informal settlements · Public space · Urban regeneration · Performance evaluation · Integrated urban projects PUI · Medellin

1 Introduction

In the twenty-first century, Medellín, the second city in Colombia, has been subject to a remarkable process of creation and renovation of public space. Public space interventions have been one of the most notable example of recent public policy and a fundamental ingredient in the city's urban transformation. They have contributed to helping the capital of the region of Antioquia become a paradigmatic example and a global benchmark for urban regeneration processes, projects and policies. In just a few years, a variety of discourses regarding rapid and efficient urban transformation have emerged, put forward by the public administration, by professionals connected to urban planning and urban design, as well as by the media and multilateral agencies. Actions in Medellín have had such a broad impact that it has been

A. Arteaga (✉)
Profesor Facultad de Arquitectura, Universidad Nacional de Colombia, Medellín, Colombia
e-mail: ajarteag@unal.edu.co

O. Carracedo García-Villalba (ed.), *Resilient Urban Regeneration in Informal Settlements in the Tropics*, Advances in 21st Century Human Settlements,
https://doi.org/10.1007/978-981-13-7307-7_5

referred to as 'the Medellín miracle',[1] and the city has come to be known as one of Latin America's 'Radical Cities'.[2] These discourses have awakened global interest in the city of Medellín, while also making it harder to talk about the particularities of the interventions or the complexity—and contradictions—of many of the processes involved.

Special attention has been paid to the urban transformation that took place in the city during the administrations of the mayors Sergio Fajardo Valderrama (2004–2007) and Alonso Salazar Jaramillo (2008–2011), whose programmes were referred to as 'Social Urbanism'. During these municipal administrations, which adopted a different focus for the use of resources in the territory, a decision was taken to intervene in the most disadvantaged sectors of the city. The process began in the north-eastern and north-western comunas, or communes, of Medellín and subsequently continued into the central eastern and central western zones, in an attempt to encompass all of the hillside territories within the city. Generally, these programmes involved four basic operations related to public infrastructure:

- The urban improvement of public space in the areas adjacent to the first Metrocable line in the north-eastern commune, opened in 1995 as a complement to the metro system. And the construction of a new cable car line connecting with the north-western commune.
- Development of urban facilities in the city, particularly the construction of educational facilities.
- The creation of library parks, designed to combine a variety of programs and uses, foreseeing their impact beyond the immediate scope of the building.
- And, finally, the overall urban planning framework, in which all these individual operations are included: the so-called Integral Urban Projects (Proyectos Urbanos Integrales, or PUIs), urban planning guidelines for intervention in the built fabric which incorporate physical, social, and institutional aspects.

The first three of these operations are easy to understand, because they are aimed at resolving specific problems and deficiencies in the areas of transport and mobility, or in the provision of facilities. However, it is worth clarifying that the proliferation of public space projects across the city can be understood as addressing a contingency and an accumulated historical debt that needed to be remedied, and they were carried out in a short period of time mainly thanks to efficient urban management. In that sense, the PUIs became a trademark of the 'Medellín laboratory'.

Likewise, it is important to note that, in recent years, a wealth of information has been published that has disseminated and notably amplified the example of Medellín. These publications, most of which have been authored by the municipal administration itself (Alcaldía de Medellín 2007, 2008, 2011a, b, c, 2014), constitute

---

[1] Maclean (2015).

[2] McGuirk (2014). The text highlights the case of Medellín along with the examples of Rio de Janeiro (Brazil), Caracas (Venezuela), Bogotá (Colombia), Tijuana (Mexico), Lima (Peru), Santiago de Chile (Chile), San Salvador (Salvador), and Buenos Aires (Argentina). These cities share the implementation of novel actions in the midst of a phenomenon inherent in the Latin American urban condition: the informal city.

the official history of the city's urban transformation. The information published to date that attempts to pinpoint the variables involved in the success of the projects all point to some of the same areas, including: the structural role of public space; the central role of the street as an essential element for transformation; the observation of everyday life as a foundation for intervention; the importance of interdisciplinary teams with a connection to the territory; the community as a fundamental agent in the process; and, finally, the fact that the process establishes a methodology that is continually updated for each new operation.

2 What Are PUIs?

Between 2004 and 2011, in an attempt at programmatic continuity, the city engaged in the government Social Urbanism project. This gave rise to the PUIs, one of the most emblematic urban planning exercises in recent years. The first of these projects, and the one that reached the highest level of development, was the urban project for the north-eastern zone (PUI Nororiental). The project had no name initially, just a site for its implementation. This PUI served as the materialization of the government's Social Urbanism programme and, from the outset, it aimed to be an 'export product'[3] and a 'replicable model':

> A PUI is a model of urban intervention that aims to improve the quality of life among residents of a specific area. To do this, it concentrates all its resources in a single territory, with the aim of focusing efforts and achieving a result that is reflected in the development and overall transformation of the local communities, in both social and physical terms. It is designed specifically to address the most depressed and marginalized areas in the city, where the State typically has a high social debt, and to be used as a replicable model of intervention' (Office of the Mayor, Medellín 2006: 27)

The most-cited definition states that:

> A PUI is a tool for urban intervention that encompasses physical, social, and institutional aspects, with the goal of addressing specific problems in a particular territory, where there has been a generalized absence of State intervention, in the interest of improving living conditions among its inhabitants. Understood as a replicable model—from previous City Hall administrations to the current one—the PUI has been executed under an accord with the Empresa de Desarrollo Urbano (EDU) [Urban Development Company] to address questions of urban and social intervention in the areas of action, attending to three main components: the physical, social, and inter-institutional orders. Having evolved into a pedagogical tool, the PUI has qualified and combined the empirical knowledge about the city held by a large number of professionals and by the citizens themselves' (EDU 2010: 47).

An Integral Urban Project (PUI) is a project with an intermediate scale, which aims to incorporate physical, social and institutional aspects, with the aim of resolving specific issues in territories that have generally been built informally, but which

---

[3]Testimony to that fact are the agents who worked on the project at a managerial and strategic level, who have became international consultants whose contractual object is the PUI.

have been consolidated and present certain common characteristics: high indexes of poverty, low human development indexes (HDI), informal urban development, low-quality public space, uncoordinated government intervention, and environmental deterioration. The goal of a PUI is to incorporate multiple development tools simultaneously, in response to a specific area of intervention—in cases where the original delimitation of the area did not depend on the traditional political divisions in the communes.

PUIs are a tool for urban intervention. Developed from outside the Colombian urban planning framework and implemented beginning in 2004, they have seen relative continuity during the last three municipal administrations. The administration of Aníbal Gaviria 2012–2015 ensured continuity in the development works of the five PUIs and committed resources for this. However, during the first year of his mandate, work teams were dismantled, and human and financial resources were channelled into new 'strategic projects'. In practice, that meant that the only PUI to be fully implemented was the first one (2004–2007).

3 Integral Urban Projects (PUIs)

From 2004 to 2011, five PUIs were developed for the city of Medellín, which, following the early success of the first project, aimed to cover basically the entirety of the city. The administration of Sergio Fajardo Valderrama formulated the North-eastern Integral Urban Project (Communes 1, 2) and the Integral Urban Project for Commune 13, while the administration of Alonso Salazar Jaramillo, formulated the Central Eastern Integral Urban Project (Communes 8, 9), the North-western Integral Urban Project (Communes 5, 6), as well as the Iguaná Integral Urban Project (Commune 7).

The first of the PUIs, the North-eastern PUI for communes 1 and 2, targeted an area of about 160 ha, including a population of about 135,000 people in 33 neighbourhoods. The main projects in this PUI encompassed: four types of mobility projects, including urban walkways, viaducts, connection paths, and manoeuvring sites; two types of public space projects based on plazas and parks; two housing and green space projects—the Juan Bobo housing consolidation and the La Herrera linear park; and, finally, three types of facilities projects—centralities, sports facilities, and mobility infrastructures. Some of the results generated by the PUI in the 2004–2007 period include 125,000 m^2 of public space, 14 new parks, four pedestrian bridges, eight pedestrian crossings, four pedestrian promenades covering a total of 2.8 linear kilometres, and a public investment of €200,000,000 (Fig. 1).

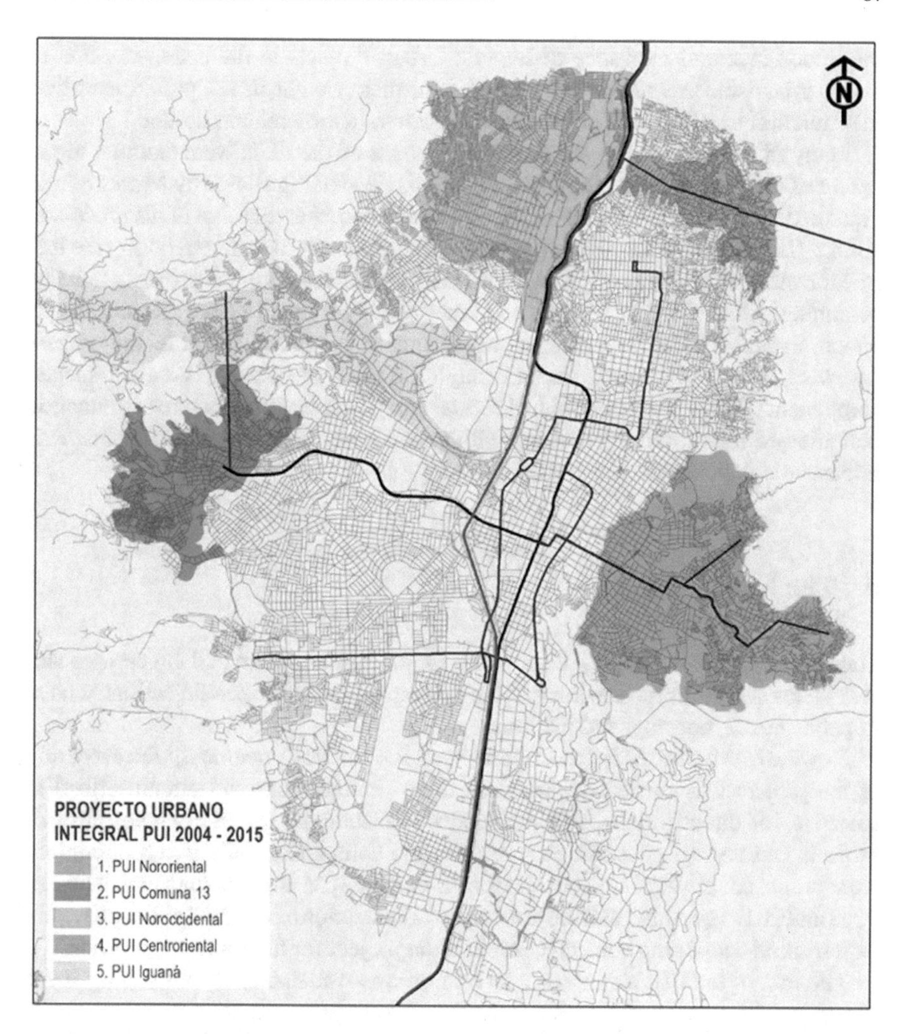

Fig. 1 Medellín's integral urban projects *Source* Armando Arteaga

4 The Case of the North-Eastern PUI

The interest of the North-eastern PUI is two-fold: on the one hand it is the result of a varied practice of interventions in the city since the 1980s; and, on the other hand, it is the starting point for a synergy between operations in the city, since four similar projects were implemented within a short span of time. However, this dynamic shifted in 2012–2015 under the administration of Aníbal Gaviria, when new strategic projects began to attract the municipality's interest, and the PUI seemed to be relegated to the background. However, the official narrative, which upholds that an urban project that is not written about may as well not have existed, recognizes the importance and relevance of Integral Urban Projects in the materialization of social urbanism. In spite of this recognition, their weight in the public narrative, their relation to precedents in the city, and their valuation remain limited.

Many of the factors that led to the weakening of the PUIs were external to the nature of the project itself. On the one hand, the leadership shown by Mayor Sergio Fajardo during the first two interventions was much greater than that of his successor, Alonso Salazar, in the remaining projects. On the other hand, whereas during the first PUI the municipal secretaries kept a low-profile role because of the project's novelty, beginning with the PUI for Commune 13 they began taking on a more prominent role, which, in practice, led to a certain polarization in the management of the project. Likewise, it is worth noting that, although the issue of housing is a fundamental component of the PUIs, it was only a central part of the first project due to a situation that allowed for combining housing with public space design in a single unit. This important component lost centrality in later projects.

5 The Master Plan

The Master Plan that includes the proposal for the North-eastern PUI is a comprehensive vision for the entire territory of the communes with 82 specific actions related to public space, housing, and facilities.

However, as indicated in the diagnosis and formulation document,[4] the execution of the project was carried out in much more limited areas and circumscribed to the zones of direct influence of the Metrocable stations, where a whole series of projects can be seen. Touching on the housing component, only two programmes were proposed: the housing consolidation along the Juan Bobo ravine; and, between 2008 and 2011 (executed outside the framework of action of the PUI, although within its territorial scope) the housing consolidation project for the La Herrera ravine.

The first of the PUI's areas of intervention was Andalucía, encompassing 107th Street and its transformation into an urban promenade, which also encompasses the interventions affecting the La Paz pedestrian bridges, to the south, and the Mirador bridge, to the north, as an extension of Avenue 48A. The middle of this new urban promenade is the site for the educational and recreational facilities, in a single block which forms the Andalucía Centrality. The endpoint of the promenade is the Andalucía Metrocable station. There are two areas that appear detached from this central system: the small "Parque de la Imaginación" [Imagination Park], at a bend in the Tirabuzón road at the intersection with Avenue 49, and the "Parque Primaveral Nuestra Señora del Camino" [Our Lady of the Road Spring Park]. On the other hand,

[4]Many of the projects identified as: "Improvement of feeder roads, local roads, and neighbourhood-scale roads" only implied changes in the direction of circulation, and others referred to as the "Construction and Refurbishment of Collective Facilities" were not carried out. In terms of housing projects, only the Juan Bobo project was in keeping with the logic of the PUI. Programmes like the "Terrace Plan" or "New buildings within city blocks" did not make it past the diagnostic stages.

the Juan Bobo housing consolidation project is located on the ravine parallel to the promenade, connecting to it via the existing road structure.

The second of the PUI's areas of intervention is known as Popular, the site for the series of projects along Avenue 42B, with the Metrocable station as the hub. To the north, and connected via the existing road, the project proposes four small parks, the "Parque del Ajedrez" [Chess Park], just across from the station, the "Parque de los Pozos" [Water Well Park], near the middle of the route, and, at the end point, the "Parque de los Lavaderos Comunitarios" [Public Washing Park] and the "Parque Pablo VI" [Pablo VI Park], which has since been replaced by new facilities. South of the promenade, there is the "Parque Vecinal Mercado Barrial" [Neighbourhood Market Park]. East of the Metrocable station, this area of intervention is complemented by the "Parque Lineal La Herrera" [La Herrera Linear Park]. Finally, facing the Andalucía station, and under the Metrocable line, there is a project that is disconnected from the Avenue 42B system, the small-scale "Parque de la Paz y la Cultura" [Peace and Culture Park].

The final area of intervention is Santo Domingo, which is centred on the section between the Metrocable station and the project for the Spain Library, which serves as the endpoint and visual point of reference for the entire promenade. Furthermore, the proposal for a loop that runs around the area along Puerto Rico Street through to the library park and then returns to the Metrocable station following the promenade on 106th Street connects three small parks: the "Parque de la Candelaria" [Candlemas Park], the "Parque de los Niños" [Children's Park] and the "Parque Mirador" [Overlook Park], as well as two neighbourhood facilities: the Santo Domingo school cafeteria and the La Candelaria school. A single project is set apart from the group of interventions: the Unidad Deportiva Granizal [Granizal Sports Centre] to the south of the promenade.

6 Creation of Public Space Through Housing Projects: The Case of Juan Bobo

As explained by Carracedo García-Villalba (2014), the process of occupation of the area of Juan Bobo began in the 1950s as an area of agricultural production, but it was not until the 1960s and 70s that its densification began, mainly as a result of migration processes from rural areas. The scarcity of land, and the lack of housing programmes and policies, made the slopes of the city of Medellín the ideal place for informal development. The regeneration project of Juan Bobo includes the neighbourhoods of Villa Niza, Andalucía and Villa del Socorro, in the northeast area of the city.[5] The 2.29 ha of the intervention area had undergone an invasive occupation process,

[5]The Juan Bobo project is located inside the North-eastern Integral Urban Project (PUI). The PUI were created by the Urban Development Enterprise (EDU) in 2004 to improve the neighbourhoods with the highest rates of poverty and informality and the least sense of belonging, and to perform several urban and social inclusion actions, based on a model of urban growth committed to the

which caused the gradual disappearance of green areas (EDU 2013). The few existing public spaces were limited to a very precarious system of paths and a series of residual natural spaces where, due to the topography, housing occupation was not possible.

Some basic principles were agreed upon between the community and the Municipality before starting the project, such as no evictions and no expropriations, but no new families settling in the territory. One of the main challenges was implementing the on-site upgrading strategy and resettling, which meant dealing with the existing community and limiting demolitions as much as possible. To achieve this objective, and with the help of the community, an infill strategy was used. This strategy consisted in the construction of buildings to relocate the affected population in the same area, along with interventions to refurbish the buildings worth to preserving, to provide public services, and to improve green areas and public spaces.

The analysis of the spatial and organization patterns of the informal settlement identified three sections with different topographic and housing characteristics, which served as the foundation for the intervention strategies. In this way, it was decided that the spaces located on the highest part of slope would be used for densification and to redefine the blocks, the middle part of the slope would be used for the redistribution of housing, and the lower slope would be used to recover the environmental space of the creek as a public space. The housing improvements were made using three strategies: new housing and on-site resettlement, which resulted in the creation of 130 new housing units, and upgrading or replacement in the remaining cases (EDU 2013).

The resulting spatial site plan is a hybrid grid based on the existing reality and adapted to the conditions of each section, without any predefined or geometrically formalized spatial pattern. By analysing the results, we observe a remarkable increase in both the outdoor space (31.7%) and the outdoor space ratio per inhabitant (1200%), the highest percentages of all cases studied. In absolute terms, 5087 m^2 of public space were built in parks, plazas, and walkways (EDU 2013), to which must be added nearly 3000 m^2 of open spaces related to environmental improvements of the neighbourhood. This increase in public space and open spaces, together with the 15% decrease of the construction footprint, resulted in residential improvements such as the decrease in housing density, the relocation of dwellings located in risk areas, the increase by 31% of the average housing living area, or the improvement and legalization of 100% of households. In the case of Juan Bobo, no appropriations of public space have been detected due to the diversity of well-defined public spaces, each with a strongly differentiated character (Carracedo 2014).

---

improvement, consolidation, regeneration, regularization, relocation, and upgrading of informal settlements.

7 Results of the PUIs and Their Impact After 10 Years. Methodology

The method followed for the assessment and analysis of the impact of public space actions carried out within the framework of the North-eastern Integral Urban Project is based on suggestions from authors such as Jacobs (1973), Gehl (2006), or Whyte (1980). It consists in using the observation of everyday practices as a fundamental basis for orienting the design of public space. Some authors have described these methods as emergent[6] or "bottom-up" systems, because of their intent to alter the traditional and hierarchical ways of designing cities. Our analysis of the cases involves three stages:

- Approach and observation: An informal and unstructured method based essentially on the observation and preparation of drawings and small field journals that established contact and allowed for familiarization with the areas under study without requiring a detailed recording process.
- Meeting: Interviews, visits and tours of the neighbourhood with social agents and community figures who facilitated the field work and helped reconstruct the unofficial history of the project, as well as understanding the community's demands, resident's participation in the urban planning process, and the experiences and opinions regarding the results of the completed projects. Also, visits, tours and field work accompanied by technicians involved in the project, which helped connect with the designs and the technical version of the project history.
- Observation and recording: Finally, records were compiled to assess the impact of the projects. A suitable tool was identified and selected to that effect: the "PLACE GAME—Place Performance Evaluation", developed by the organization Project for Public Space (PPS).

The system proposed for our study was based on this tool, designed to help evaluate a space through the recognition of four fundamental attributes. It was complemented by contributions from Jane Jacobs and Jan Gehl, as well as by the specific local contributions describing the operations that took place in the territory of the North-eastern PUI beginning in 2004.

8 Assessment Matrix: Criteria for the Evaluation of Public Space Projects

The success or failure of the Integral Urban Projects, and in this particular case the public space projects and facilities that they proposed, cannot be attributed to a single factor. The assessment methodology proposed for this study takes into account three components with an interconnected interpretation:

[6]Johnson (2001).

- The environment (characteristics of the place and its dynamics): based on the recognition of factors that generate diversity, which are closely related to the specific occupation dynamics of each area.
- Design decisions (criteria): they aim to highlight the criteria underlying the interventions in order to explain some successes and failures in the projects and in the areas of intervention.
- Performance of the site (behaviour): the information collected during the periodic visits provides an average of the impact of the projects and their daily behaviour based on the four fundamental attributes. In this case, it is a qualitative assessment, but it also provides quantitative indicators, which can be organized into areas of intervention and types of projects.

9 On the Current State of the Projects

Figure 2 "Performance Evaluation North-eastern PUI Projects" shows the results of the evaluation of 23 public space projects and facilities in the area of the North-eastern PUI. These projects represent the ones that were finished, out of the total of 83 operations that were proposed for the PUIs and identified at the beginning of this article. The graphic provides an idea of the location and intervention in the territory, the individual impact of each project, and the overall impact in each area of intervention. Likewise, the graphic helps to identify, using three different colours, the status, impact, and relevance of the 23 projects 10 years after their implementation. Consequently, we can derive the following conclusions from the analysis and evaluation (Table 1).

Three projects, identified with green circles, are recorded as important and significant, in good condition, and with a high impact (more than 75%). Ten projects, identified with yellow circles, show an intermediate impact, tending toward the top category (between 50 and 75%). Finally, 10 projects are identified with a red circle, indicating a level of performance and impact of less than 50%; these are often sites that have been abandoned or are not very relevant in the daily lives of neighbourhood residents.

Within this global interpretation, it is worth highlighting that nine projects appear as important and significant. On the one hand, three of them are related to facilities—Unidad Deportiva Granizal, Centralidad Andalucía, and Cedezo—which serve three fundamental needs of the local population: sport; a meeting place; and access to a variety of opportunities in education, culture, and employment. On the other hand, six of the projects are related to interventions involving streets, such as the 107th Street urban promenade, the Andalucía Centrality, the Overlook Bridge, the improvement of Avenue 42B, the urbanisation of Puerto Rico Street, and finally the 106th Street urban promenade that ends in the Children's Park. All these projects that affect local mobility lead to improved community living and social integration, as well as greater comfort in the use of public space. As such, they are in good condition and their impact on the neighbourhood population is generally high.

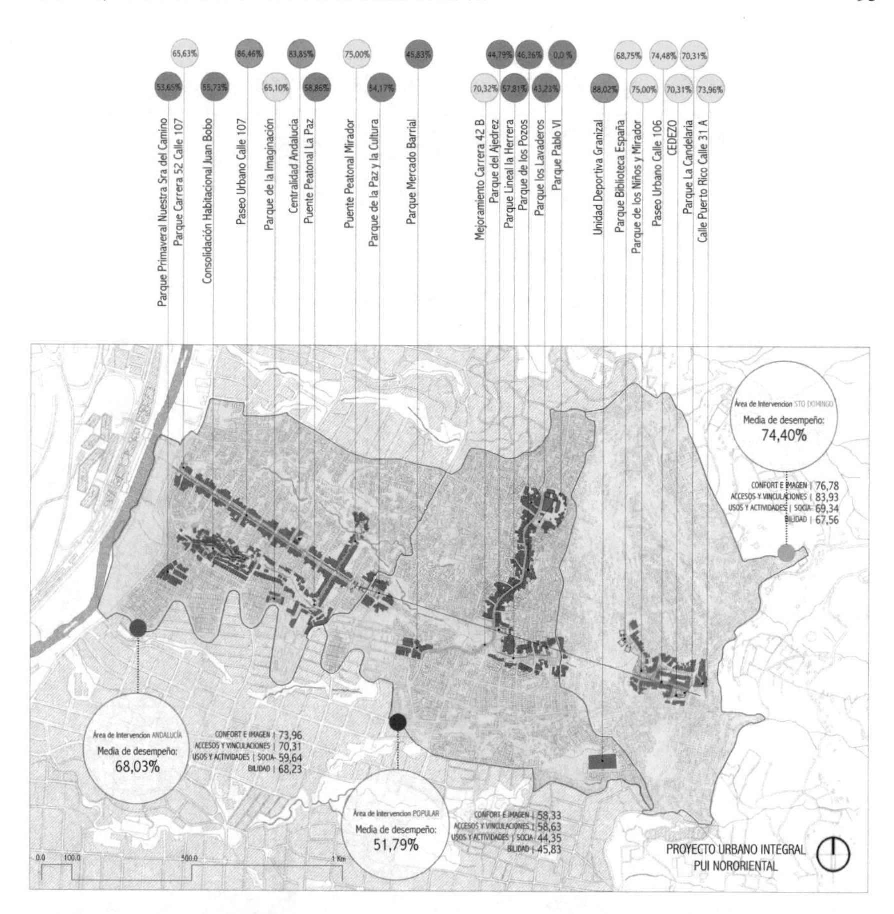

Fig. 2 Performance evaluation North-eastern PUI projects. Circles from left to right: Andalucia intervention area: average performance 68.03%/comfort and image: 73.96/access and links: 70.31/uses and activities: 59.64/sociability: 68.23. Popular intervention area: average performance 51.79%/comfort and image: 58.33/access and links: 58.63/uses and activities: 44.35/sociability: 45.83. Sto Domingo intervention area: average performance 74.40%/comfort and image: 76.78/access and links: 83.93/uses and activities: 69.34/sociability: 67.56 *Source* Arteaga 2016

Three of the projects have fallen into an advanced state of deterioration. These are: the Imagination Park, the Chess Park, and the Peace and Culture Park. In the case of the latter two, the deterioration has not resulted from systematic and intensive use on the part of the population[7]; the deterioration is associated with abandonment due to lack of use. This situation contrasts sharply with the situation of the public space projects associated with the implementation of the Metrocable K line. Although the small squares bordering on the Metrocable stations, (although they are directly adjacent to the Chess Park and the Peace and Culture Park) show both limited use

[7]This can be interpreted as a success in technical and material aspects of the project, but as a failure in its use.

Table 1 Criteria for the evaluation of public space projects

CRITERIA FOR THE ASSESSMENT OF PUBLIC SPACE PROJECTS

1. TECHNICAL INFORMATION		2. PHOTOGRAPHIC RECORD		3. ASSESSMENT	
	Design group		Before / 2004		Description
	Social work - Communication				
	Technical advisory		2007		
	Project execution		Status 2015		Actual state
	General project data				

4. CHARACTERISTICS OF THE ENVIRONMENT	TYPE OF SETTLEMENT	PARCELATION	ROAD LAYOUT	STREET TYPE
	Formal and private urbanization	Planned	Lineal	Urban arteries
				Continuity streets
	Pirate development	Organic	Net	
		Mixed		Neighborhood distribution streets
	Invasions	Invasiva	Arboreal	Neighborhood streets
				Roads and trails

5. DIVERSITY GENERATING FACTORS			
Combination of primary uses	LOW	HALF	HIGH
Small blocks		NO	YES
Mix and presence of old buildings	LOW	HALF	HIGH
Concentration density	LOW	HALF	HIGH

6. REQUIREMENTS TO PROJECT CONTACT		
Walls	WITH	WITHOUT
Distancces	LONG	SHORT
Speeds	HIGH	DOWN
Levels	VARIOUS	ONE
Orientation	BACKWARDS	HEAD ON

7. PERFORMANCE EVALUATION		1	2	3	4
Comfort & Image	Overall attractiveness Feeling of safety Cleanliness / Quality of Maintenance Comfort of places to sit				
Access & Linkages	Visibility from a distance Easy in walking to the place Transit access Clarity of information / Signage				
Uses & Activities	Mix of stores / services Frequency of community events / activities Overall busy-ness of area Economic vitality				
Sociability	Number of people in groups Evidence of volunteerism Sense of pride and ownership Presence of children and seniors				

Source Arteaga (2016)

and limited social appropriation, they are well maintained and cared by the concessionaire that runs the Metrocable service. Whereas these small spaces have been modernized, which makes their "age" difficult to identify, the rest of PUI projects and infrastructures exhibit premature aging because responsibility for their management and maintenance falls exclusive to the community, whose resources are limited resources.

Finally, despite their good physical condition, five of the projects are of little relevance and impact on the local population due to their limited use. In this case, we are referring to the Water Well Park, Our Lady of the Road Spring Park, Candlemas Park, the Neighbourhood Market Park, and the park at Avenue 57 and 107th Street. Setting aside the fact that three of the projects are undergoing a process of reconstruction or total renovation (the Guadua Bridge, under reconstruction after its collapse in 2013, as well as Pablo VI Park and the Spain Library Park, under refurbishment) and assuming that their situation is remediable, we find that 30% of the projects, eight of the 23 projects, have failed to withstand the test of time—in questions of daily use in the former case, and in technical aspects in the latter.

With regard to the careful observation of everyday life during the process of designing the projects as a guarantee of high-impact projects, the results are poor. The general state of the projects 10 years after their implementation indicates a baseline to take into account in the development of public policies.

10 Lessons Learned from the Experience of Using PUIs as Tools for the Regeneration of Informal Neighbourhoods

PUIs represent a new local tool for addressing marginal territories which, unlike the exercises initiated in the 1980s, were implemented under a new political and regulatory context. In Medellín, the approach to the intervention in informal settlements characterized by a high degree of insecurity already had an important history: 20 years of practices that preceded the first PUI. The experiments carried out in 1983 and 1987 through the urban rehabilitation program in the Medellín garbage dump, known as "Moravia", demonstrated how to approach complex informal territories characterized by poverty and insecurity. At the beginning of the 1990s, the work of the Consejería Presidencial para Medellín y su Área Metropolitana (CPM) [Presidential Council for Medellín and its Metropolitan Area], began a government effort to address the ongoing social crisis, making Medellín into a national issue. The Programa Integral de Mejoramiento de Barrios Subnormales en Medellín (PRIMED) [Integral Substandard Neighbourhood Improvement Program in Medellín] developed methodologies for the intervention and treatment of at-risk settlements, mainly in its first phase from 1991 to 1996. Also, in the same period, the program Núcleo de Vida Ciudadana (NVC) [Hubs of Civic Life], proposed a new role for community facilities. In 2002, convention 017, the Propuesta de Regularización y Legalización Urbanística Zona Nororiental de Medellín [Proposal for the Regularization and Legalization of North-eastern Medellin] presented an important updated diagnosis of the most vulnerable and priority areas. Finally, in 2003, an undergraduate thesis in architecture presented an idea for an intervention in the area of influence of the city's first Metrocable station in the Northeast commune. All these experiences as a whole, more than points of reference, were the raw materials for the PUI and the directory of professionals to be involved.

All of these prior experiences, combined with broader and better management tools, a greater capacity for action given the attributes of the managers involved, more resources allocated to public works, and a period characterized by synergy among the agents involved and their demonstrated leadership, made it possible for ideas (previously explored), rules (that had to be, or should have been, invented) and resources (now surpluses) to serve as the foundation for the design of a large number of urban projects in a short amount of time.

Although the gradual recognition of the PUI's impact as a tool never occurred, and similar operations were devalued along the way, it is important to highlight the achievements of the PUI as a tool, as well as some remaining challenges[8]:

- In the local context that emerged outside the Colombian urban planning framework, the PUI constitutes a significant contribution in the design and execution of urban renewal policies that attempt to incorporate physical, institutional and social components within a single operation.
- The physical component refers to construction, mainly public space and facilities; the institutional component involves inter-institutional and inter-sectoral coordination to work together; and the social component, although it is more difficult to assess, makes reference to the strengthening of community organizations and the impact of assistance programs that emerge. In principle, the project needs to be legitimized through community participation. Although what that means is still being consolidated, there is evidence of an emphasis on defining key agents and stages in the process. The approach based on coordinating the different stakeholders has been consolidated as a fundamental option in urban renewal.
- The management model for the first PUI was not formalized under an institutional framework. In, fact it never became an official policy for intervention, and what had initially been an advantage in developing the first PUI—an unusual amount of power, leadership and manoeuvrability, which created an exceptional synergy between the departments that were involved—became an argument against it in the development of the later projects. Leadership began to falter during the Salazar administration (2008–2011) and the program gradually disappeared during the Gaviria administration (2012–2015). This "deinstitutionalization" became evident in later similar projects, leading to a dilution of their relevance over time. In that sense, the role of institutions is important and needs to be strong and consistent, open to dialogue and inclusive.
- One factor that actually worked against the later PUIs was, contradictorily, the early success of the first project. The eagerness to generate new projects overshadowed the knowledge that had been acquired during the experience in the north-eastern commune. The process of involving people in the project—which required an intense, continuous, systematic preliminary process of understanding the territory, walking through it, learning about it, diagnosing it in detail, and formulating in situ strategies that would be precise and coherent with the territory and the community—set aside and forgotten in the applying the method to new cases. This ended up having a clear and direct effect on the impact and results of the new projects that were implemented. In that sense, it is important to have

[8]The new administration of Medellín, under the mayor's office of Federico Gutiérrez (2016–2019), has expressed its intention to resume Integral Urban Projects as a strategy for intervention in the city.

detailed knowledge of previous experiences and of the particularities of each territory. While successful processes are an essential foundation for generating new projects, it is equally important to adapt them to the existing reality in each case, as opposed to subscribing to the idea that "one solution fits all".

- Another mistake in extrapolating the experience from the first PUI to four other strategic areas of Medellín was ignoring that, conceptually and originally, their area of action should not be dictated by traditional political-administrative divisions. The unnecessary extension of operations to adhere to administrative political boundaries implied, in practice, taking on unknown territories, without being able to rely on the contributions of the north-eastern precedent. The new exercises were unable to span the sheer scale of the operation. One of the greatest legacies of the North-eastern PUI was that it was derived from the size of the neighbourhoods. The scale of subsequent experiments was excessive, disproportionate and uncontrollable from the point of view of urban management; this is another aspect to be taken into account in the implementation of urban renewal projects in informal areas.
- Housing as a fundamental component of the PUI was only a factor in the first experiment, and it was the result of administrative circumstances that permitted a single municipal entity to design the housing and surrounding public space. This approach may seem obvious, but it was impossible given the budgetary and administrative vision at the time. Where operations of this nature are carried out without the housing component, the results will always be modest and low impact. In that sense, any urban renewal operation should incorporate housing along with the infrastructure and public space components.
- The sustainability of the project revolved around the idea that the maintenance and management of the finished project should be the exclusive responsibility of the community. Although this proposal may be empowering to the community, it is not a guarantee of its sustainability in the short, medium or long term. As such, we would posit that the intervention cannot be qualified as "finalized", despite what the administration seems to believe. Sergio Fajardo's campaign slogan when he ran for governor of Antioquia (2012–2015) was: "We did it in Medellín, and we'll do it for all of Antioquia", implying that the work in Medellín was done, when, in fact, it had only just begun. The precedent of the experiment in Moravia, during the 1980s, where the management and sustainability of the projects fell prematurely and exclusively to the community, and where the municipal administration left the area assuming that their role was over, ultimately resulted in a rapid process of deterioration of the area in general, to such an extent that the municipal administration had to return to address it 20 years later with a new public rehabilitation effort. The current state of the North-eastern PUI projects shows that they have followed the same path and, a decade after their implementation, it is evident that a new public intervention to address these projects will be necessary. We need to find mechanisms of empowerment and progressive transfer to the communities combined with pedagogical policies to help residents take charge of their local public space and infrastructure. A feeling of belonging can help

toward the maintenance of projects and spaces, which, together with management policies and economic support, can foster use and maintenance from within the neighbourhood itself.

References

Alcaldía De Medellín (2007) Del miedo a la esperanza, testimonio gráfico. Taller de la edición, Medellín

Alcaldía De Medellín (2008) Transformación de una ciudad. Mesa Editores, Medellín

Alcaldía De Medellín (2011a) Laboratorio Medellín. Catálogo de diez prácticas vivas. Mesa Editores, Medellín

Alcaldía De Medellín (2011b) Medellín, guía de la transformación ciudadana 2004–2011. Mesa Editores, Medellín

Alcaldía De Medellín (2011c) Bio 2030 Plan director Medellín. Valle de Aburrá. Mesa Editores, Medellín

Alcaldía De Medellín; Isvimed; Complexus (2014) Carta Medellín. Solingraf, Medellín

Arteaga A (2016) Medellín: espacio público re-potenciado. Caso de estudio: Proyecto Urbano Integral (pui) Nororiental (2004–2007). Tesis doctoral, Universidad Politécnica de Cataluña (UPC), Escuela Técnica Superior de Arquitectura de Barcelona

Carracedo García-Villalba O (2014) The form behind the informal. Spatial patterns and street-based upgrading in revitalizing informal and low-income area. In: Heng CK, García-Villalba OC, Ye Z (eds) Asian urban places. Great Asian streets. Publisher CASA Centre of Advanced Studies in Architecture. National University of Singapore

Dávila J (compilador) (2012) Movilidad urbana & pobreza. DPU, UCL, Universidad Nacional de Colombia, Medellín

Empresa De Desarrollo Urbano (2013) Medellín. Modelo de transformación urbana. Proyecto Urbano Integral-PUI en la zona nororiental. EAFIT, AFC, EDU, Alcaldía de Medellín. Empresa de Desarrollo Urbano EDU: "Los proyectos urbanos integrales". EDU Medellín. Alcaldía de Medellín

Empresa De Desarrollo Urbano; Departamento Administrativo De Planeación (2005) Convenio No. 4800000316 de 2004 - Informe final PUI Nororiental etapa diagnóstico y formulación. Medellín, EDU

Empresa De Desarrollo Urbano; Oopp (2010) Libro blanco (inédito). Medellín

Empresa De Desarrollo Urbano; Universidad Eafit; Agencia Francesa De Desarrollo (2012) Medellín Modelo de transformación urbana. (inédito). Medellín

Empresa De Desarrollo Urbano; Banco Interamericano De Desarrollo; Alcaldía De Medellín (2014) Equidad territorial en Medellín. Mesa Editores, Medellín

Gehl J (2006) La humanización del espacio urbano: la vida social entre los edificios. Reverté, Barcelona (1 edición en ingles 1971)

Gehl J (2010) Cities for people. Island press, Washington

Gehl J (2013) How to study public life. Island Press, Washington

Jacobs J (1973) Muerte y vida de las grandes ciudades. Península, Madrid

Johnson S (2001) Emergence: the connected lives of ants, brains, cities, and software. Scribner, New York

Maclean K (2015) Social urbanism and the politics of violence: the Medellín miracle. Palgrave Pivot, London

McGuirk J (2014) Radical cities: across Latin America in search of a new architecture. Verso, London

Mcguirk J (2015) Ciudades radicales. Un viaje a la nueva arquitectura latinoamericana. Turner, Madrid
Whyte W (1980) The social life of small urban spaces. Project for Public Spaces, New York

Education and Pedagogy. Urban Regeneration Through Training

Improving opportunities for access to decent and formal employment through training, involves the development of vocational training programs identified demand from both labour and output, the implementation of transfer programs conditioned to schooling, and the production of facilities for a broad learning and social development. Usually, the level of education is inversely related to the level of informality in the labour market.

Capacity Building and Education in Latin America. Contributions and Limitations of the Venezuelan and Colombian Experiences, Lessons Looking into the Future

David Gouverneur

Abstract Capacity building is needed the most in creating widespread awareness, and training those that will be the facilitators of the regeneration process of informal settlements. As per UN-Habitat's estimates, 2 billion inhabitants (For detailed statistic on sources and types of violence in informal settlements, see United Nations in The challenge of slums. Global report on human settlements. Official, Earthscan Publications Ltd., London 2003), double that of today, will be self-constructing their habitats and creating new informal settlements occupying new territories. This represents a major environmental, socio-economic and political challenge if left to grow spontaneously (For issues derived from urban inequalities, see Grauer in Democracy in Latin America. David Rockefeller Center for Latin American Studies, Harvard University, 2002). Learning from the Latin American experience, where informality is the principal form of urbanization, planning laws have been enacted or revised, revealing the importance of addressing informal urbanization, which requires the proactive involvement of the public sector. This gave rise to the phenomenon where innovative planning, design and managerial paradigms have been brought in to encourage community engagement. This approach is explained in this chapter with reference to two emblematic Latin American case studies—Caracas in Venezuela and Medellín in Colombia—that focus on upgrading or rehabilitation of informal settlements, pointing out their achievements and limitations, revealing the importance of working on the ground, hand in hand with the communities, and the need for committed political support and efficient municipal management. The lessons derived from these pioneering approachs reveal a series of insights: the importance of changing paradigms and enacting policies to provide skills at the national scale; the need for committed political and technical actions that can tackle informal settlements at the metropolitan scale on a par with site-specific initiatives; and the need for community engagement and the indispensable role of municipal governments working in tandem, and on site, with the communities. The article concludes by providing innovative planning and design solutions for emerging informal settlements in the form of flexible design components, adaptable to different

D. Gouverneur (✉)
Stuart Weitzman School of Design, University of Pennsylvania, Philadelphia, USA
e-mail: dgouverneur@gmail.com

O. Carracedo García-Villalba (ed.), *Resilient Urban Regeneration in Informal Settlements in the Tropics*, Advances in 21st Century Human Settlements,
https://doi.org/10.1007/978-981-13-7307-7_6

site conditions, which will provide predominately informal cities with higher levels of performance and help residents attain decent living conditions.

Keywords Self-constructed settlements · Reducing inequalities · Informal armatures · Hybrid approach · Capacity building

1 Introduction

Capacity building is a rather ample term, which usually refers to the process of providing criteria, skills, tools, and actions that can effectively improve the performance of the public sector, institutions, NGOs and communities, thus helping neighbourhoods to thrive and adapt to changing conditions. In order to do so, and according to the characteristics of each context, the capacity building framework requires addressing multiple aspects simultaneously, which are the main focus of other chapters in this publication.

For instance, if the academic sector cannot undertake research or train professionals to acquire the tools to respond to the demands of the informal city, little will be achieved even if there is political will, and even if the legal systems have been adjusted to recognize informality as a main component of cities, and vice versa. Similarly, if planning paradigms and implementation tools are not compatible or ignore the needs of self-constructed or informal neighbourhoods, there will be limited results, even if academia has trained professionals with adequate skills or if efforts have been made in terms of infrastructural moves and spatial connectivity.

Therefore, the central topics of this publication necessarily overlap with the means of capacity building. However, this chapter focuses on the nonphysical aspects that seem to be more directly related to the notion of capacity building, commenting on the difference between urban renewal versus urban regeneration, changes to planning and design paradigms, educational contributions, as well as the role of political, institutional and community engagement, and partnerships as pivotal components in attaining success when dealing with the improvement of the living conditions in informal settlements.

It is important to note that the academic milieu frequently views informal urbanism with an almost romantic and anthropological fascination, praising only the creativity, dynamism, and resiliency of residents, in addition to the adaptability, complexity, and diversity of activities and uses, and the morphological and aesthetic attributes of these organic piecemeal settlements. This biased take leads to a hands-off or, at best, light-touch attitude towards the informal.

While these attributes are certainly present and desirable, in most cities of the Global South—where the informal is already the dominant form of urbanization and, in many cases, results in some of the largest urban agglomerations in history—the informal city requires urgent attention and intervention. The levels of violence, environmental hazards and sanitation problems, poverty, isolation, commute times for residents to access better jobs and services in the formal city, and the loss of life and assets resulting from settling in high-risk locations, compromises the welfare of these communities.

These harsh conditions lead to social unrest, affecting the performance of the larger urban systems and even the governance of cities or entire nations. The challenge seems to lie, then, in determining how to act upon aspects that informal settlements cannot attain on their own, while letting the intrinsic positive aspects of informality thrive. The case studies included in this article speak about this delicate balance.

2 On Urban Renewal and Planning/Design Paradigms

In the American continent, and particularly in Latin America, the term urban renewal usually refers to the violent demolition of an existing urban fabric accompanied by social displacement. That is not the essence of this book, which is centred on the improvement of informal settlements and low-income communities.

Therefore, the terms urban regeneration, upgrading, rehabilitation, or revitalization of informal settlements, and their integration into larger urban systems, are a better fit when it comes to describing the current state of the art, policies and programs carried out in this region, and on other continents. Additionally, as we will explain, intervening in the self-constructed city requires operating surgically and in close proximity with local actors, while responding to contextual and cultural nuances.

It is important to note that, during the early half of the twentieth century, Latin American countries like Venezuela, Brazil, and Mexico experienced rapid processes of industrialization, modernization, and economic growth; it was a period in which urban renewal, as we described it above, was considered as the mainstream approach by leading academic institutions worldwide, influencing policy making and professional practice, and resulting in tabula-rasa urban interventions in both formal and informal fabrics.

Such ideas and operations represented the materialization of modern urban paradigms, which sustained that the traditional, piecemeal, and self-constructed city was intrinsically sick, and thus could not respond to contemporary demands, frequently leading to the razing of old urban cores and to the development of radically different urban conditions.

Urban renewal was seen as the opportunity for testing out the modern planning and design principles, which posited: road and vehicular efficiency, single-use district zoning, the elimination of street walls, and the deployment of detached (objects in the park) buildings, all framed as an efficient machine-oriented take on the city, fuelled by the real estate market and, in some cases, by formal social housing policies.

The urban results were rather standardized, not culturally grounded or site-specific, and were of a top-down nature, generally carried out by central governments with no regard for community participation. Under this approach, entire traditional city centres and neighbourhoods were demolished, disrupting socio-economic systems, severing urban grids and connections, which affected the performance of cities for decades. The paradigms embedded in the Modern Movement, which were widely used in developing countries, also fostered rapid city expansion, resulting in fragmentation, sprawl, and social segregation.

While such planned urban transformations were taking place, these thriving cities were experiencing the growth of informal settlements, as migrants from rural areas and from other nations arrived in search of jobs, services and amenities, or fleeing from poverty and/or violence in the countryside. In many ways, the emerging self-constructed conglomerates reproduced the conditions that the Modern Movement was attempting to eliminate through urban renewal operations, including: granularity, intricacy, mixed-use and pedestrian-driven environments, piecemeal and transformative design conditions, etc.

In other words, the predominant modern planning, design, and managerial paradigms ignored or rejected the informal city, with even worse consequences, since real-estate-driven planning paradigms were pushing the populations with more limited economic resources out of the urban perimeters, as defined in the planning instruments, onto peripheral and often challenging risk-prone sites—which, in turn, did not offer access to jobs or to services and infrastructure. In other words, the informal occupation of non-urbanized land was the only option for those with fewer resources who cannot access the formal real estate market, lacking formal jobs or savings.

3 On Lessons of Holistic Visions and Programmes for the Revitalization of Informal Settlements, and the Role of the Political Sector

It is important to note that towards the end of the twentieth century, in many Latin American nations more than half of the urban population already lived in informal areas. Having been neglected for decades, as the settlements grew in population and gained political leverage, they began to receive the attention of governments, initially with rather simple moves like providing them with potable water, electricity, basic infrastructure, paving paths and public staircases, schools, medical and sport facilities. These top-down actions, while very relevant for the communities, did not alter the marginalized, segregated, and stigmatized nature of these neighbourhoods, nor did they provide opportunities in terms of jobs, services, and amenities offered by the formal city.

During this period of early interventions in informal neighbourhoods, the academic and professional milieus lagged behind, as most Latin American universities and professionals ignored the informal city, even contributing to spreading cultural/professional biases against them. In this early stage, it was the political sector, with limited vision and technical tools that initially was paying attention to a growing and complex urban phenomenon. This embryonic public attention, frequently carried out during pre-electoral periods, was aimed at providing electricity, potable water, paved pedestrian paths and stairs in very steep settlements, and some minor sport or educational facilities. In some cases, the politicians looked the other way, or even encouraged squatting in order to gain votes, not considering the risks or limitations of where the occupations were taking place.

However, as informality became the predominant form of urbanization in many Latin American nations, planning laws were enacted or revised, revealing the importance of addressing informal urbanization. This required proactive involvement from the public sector, although the technical and managerial tools at their disposal were limited initially. It was the dawn of new planning, design, and performative paradigms for the self-constructed city.

Conversely, the past decades have been characterized by cutting-edge practical experiences in different South American countries, conceived as holistic regeneration/rehabilitation projects. Valuable information, working methods, and design criteria are now being shared and adapted to various contexts on an academic, institutional, and political level. While various academic explorations and research efforts had been undertaken in some Latin American countries to address informality and self-construction—mainly in Venezuela and Peru, the practical experiences, which allowed for testing solutions on the ground, measuring achievements and failures, and opening up new directions, fed back into the academic world and served as points of reference for further on-site actions.

There are notable contributions in Venezuela, Brazil, Colombia, Bolivia, listed here in chronological order, which allowed those countries to learn from each other.[1] However, it is worth pointing out that there are no dogmatic solutions when it comes to dealing with the improvement of informal settlements and with their spatial and performative integration into the formal city. Each country, region, city, and neighbourhood is different and thus requires a tailored approach.

The case studies we will be discussing here refer to two emblematic Latin- American projects focused on the upgrade or rehabilitation of informal settlements, with which we are particularly familiar. We will point out their achievements and limitations, revealing the importance of working on the ground, hand-in-hand with the communities, along with the need for committed political support and efficient municipal management.

The case studies are:

- The plans and competitions for Caracas and other cities in Venezuela, a country that pioneered the research and elaboration of plans of this nature in the region, but where there was ultimately limited success in terms of practical results, and
- The case study of Medellín, Colombia, a city that was once considered the most dangerous city in the word (in the 1980s–1990s), which has become a world reference in terms of the successful urban transformation of challenged low-income communities and citywide performance.

[1]For more information on what makes these cases particularly relevant in the context of working with informal settlement improvements, see Gouverneur (2014).

4 The Venezuelan Experience

In the mid-1990s, I occupied the position of National Director of Urban Development of Venezuela. During that time, from this public position, I supported the work of a group of talented researchers, including Teolinda Bolívar, Josefina Baldó and Federico Villanueva from the School of Architecture and Urban Planning of the Universidad Central de Venezuela. These researchers were perhaps the first in the region to warn of the consequences of inaction in relation to informality. For many years, their work was ignored or mocked, even in public, at the Schools of Architecture and City Planning, since their findings were not considered interesting by the status quo from a design perspective, at a time when close to 45% of the country's population lived in informal settlements.

The Metropolitan Plan for the Improvement of Informal Settlements in Caracas, carried out under their leadership for the Ministry of Urban Development, where I acted as General Director of Urban Planning, dealt with an array of neighbourhoods, encompassing 1.3 million inhabitants—a very high percentage of this capital city's population of 3.5 million.[2] Their methodology proposed working at two scales, identifying what were referred to as UPUs (Urban Planning Units) and UDUs (Urban Design Units).

The UPUs addressed large-scale metropolitan challenges, such as quantifying the number of inhabitants living in settlements situated in high-risk areas that would need to be subject to relocation strategies, as well as major investments in relation to water supply, sanitation and mobility systems at a metropolitan/city scale. These were operations that would fall under the responsibility of federal and regional agencies, due to their systemic nature, large scale, and high costs, which would be difficult to tackle by the different municipalities that conformed the Metropolitan Area.

The UDUs focused on the specificities of each neighbourhood, such as requirements in terms of public spaces, accessibility and transportation, community infrastructure, services, and also additional demands for relocation or substitution of housing in order to secure the spatial requirements to achieve these site-specific interventions. This set of initiatives, while receiving support from the national and regional governments, would necessarily require a municipal proactive response, working closely with the residents of each neighbourhood.

In less than a year, the team produced a large amount of quantitative and qualitative technical information, derived from their previous research but updated and verified by quick on-site observations with the participation of experts in different fields: hydrology, geology, seismic studies, infrastructure, transportation, etc., and also from experts in the social sciences. These studies provided important data for the formulation of a comprehensive program, which, due to the magnitude of the task in terms of urban area, population, and the configuration of the majority of the

[2]For additional information on the plan, see Josefina Baldó, and Teolinda Bolívar. La cuestión de los barrios. Caracas: Fundación Polar, Universidad Central de Venezuela, Monte Ávila Editores, 1996.

settlements located on a very steep terrain (with virtually no infrastructure, mobility systems, public spaces or services), would have to be carried out over a 15-year period.

The relevant information derived from this plan included the percentage of informal settlements and population that was effectively located in high-risk areas and thus had to be relocated in substitution housing—preferably to be built in the vicinity of the areas where they lived. This resulted in approximately 80,000 inhabitants subject to relocation programs. This data was important, since it revealed the inaccurate projections of those who argued that almost the entire percentage of those residing in these settlements occupied unstable, flood-prone land or other high-risk locations, and therefore the only solution was the total demolition of the informal settlements. These studies also demonstrated that another 70,000–75,000 inhabitants had to be relocated in order to carve out the spaces required to provide basic accessibility—such as small service roads in areas where the residents would frequently have to walk up or down over 100 m of grade elevation from the nearest access roads and public transportation—or to create public spaces and incorporate community services and amenities.

Both numbers combined represented about 15% of the total population, a figure that could be satisfactorily managed in order to provide a substantial improvement of living conditions in these areas. Additionally, the plan made it possible to estimate general implementation costs, roughly US$1 billion, and to determine how these interventions should be phased, how different institutional actors (federal, regional and local) would interact, and the means of engaging with communal organizations.

When President Hugo Chávez assumed his first mandate, Professor Baldó was named the President of the Venezuelan Housing Council (CONAVI). She made plans to set the improvement of informal settlements as a national priority, instead of the construction of new subsidized housing, which, as had been proven, was always out of reach for the less affluent and most populous groups. Consequently, CONAVI organized hundreds of competitions nation-wide, based on the general criteria and implementation strategies derived from the methodology of the Caracas Plan,[3] adapted to site-specific conditions. An important lesson from this experience is that all those interested in participating in the competitions were required to register cross-disciplinary teams and also to take a two-week flash course on the nuances of informal settlement habilitation, since there was little expertise on the subject and on this type of teamwork in the country, despite the high numbers of highly qualified local professionals in related fields such as planning, architecture, urban design, infrastructure, sociology, etc.

The teams that won the competitions were granted the responsibility of developing the projects, working closely with the neighbourhood residents. This requirement was referred to as "acompañamiento social" (or social engagement) and included defining which pilot projects would respond to their communities' priorities and could also serve as catalysts for change. These actions would be accomplished within a very

[3]Ibid note 4.

short time-frame in order to enhance trust and credibility among public agencies, technical teams and residents.

This initiative represented a major change of paradigms, shifting efforts and funding from the construction of subsidized formal housing (backed by developers), which only reached a more affluent segment of the population, to the holistic improvement of the self-constructed urban habitats, where close to half of the less affluent inhabitants of the city resided. Consequently, this initiative had a major impact on technical and community-oriented capacity building on a national scale, created a pro-informal settlement scenario, and strongly influenced academia and professional practices. This methodology is recommended for countries that are beginning to pay attention to informal settlements in national policy, and where there is little academic and professional expertise on the topic, since it allows for rapidly diffusing knowledge and producing quick responses.

The Venezuelan experience, however, did not have a happy ending. Despite the government's "socially oriented" policies, it promptly adopted a highly centralized and militarized agenda, displacing the professional segment from leadership positions and weakening the role of municipal governments. The influential military sector had a persistent bias against initiatives focused on informal settlement improvements and returned to the lucrative business of building social housing, in many cases on land in the urban peripheries, detached from the cities, and even within military camps. This brought the Venezuelan informal settlement program to a virtual halt, with the consequent frustrations of millions of residents who had believed in its benefits and had engaged in its conception.

While one can argue that the neglect of these large self-constructed areas for many decades (from the 1950s to the late 1990s) led to social resentment and frustrations, which sparked the emergence of new political forces that resulted in the so-called Socialist Revolution in Venezuela at the dawn of the new millennium, it also derived over time into a harsh militarized, criminal and drug-related dictatorship. The regime's rhetoric of "helping the poor" without effective economic reforms, or attention to the territorial/urban, physical and performative conditions of the settlements also explains the severe deterioration of living conditions and violence that nowadays affect the majority of the population in this country.

The lessons derived from this pioneering approach reveal: the importance of changing paradigms and policies to provide skills on a national scale; the need for committed political and technical actions that can tackle informal settlements on the metropolitan scale, on a par with site-specific initiatives; and the need for community engagement and the fundamental role of municipal governments working in tandem with the communities on site. The failure of the Venezuelan experience can be contrasted with the success achieved in Medellín, a city with striking similarities to Caracas in terms of geographical setting—a narrow valley at 1200 m of elevation, a population of 3 million people, and with an urban configuration of self-constructed settlements in which the most challenging ones occupy very steep slopes, making the comparison even more relevant.

5 The Medellín (Colombia) Experience

Medellín offers perhaps the most notable example of informal settlement improvement operations, a world-renowned and many times award-winning[4] effort envisioned as part of a wider urban strategy. Medellín stands out for the quality of the design's management skills and the impact it had on the communities and the city at large. Only two decades ago, Medellín ranked as the most violent city in the world. Statistics revealed that the highest percentage of the violent crime occurred in the informal areas located on the almost inaccessible mountaintops. In April 2014, the city hosted the World Urban Forum and participants from all over the world were able to appreciate how creative design and transparency in the management of urban change could improve living conditions in one of the city's most distressed areas.

The architect Alejandro Echeverri, a pivotal technical, managerial, and pedagogical key figure in the city's transformation process during the administration of Mayor Sergio Fajardo, and now the Director of URBAM (the Centre for Urban and Environmental Studies and professional practice at EAFIT University in Medellín), always refers to these interventions as projects, and not as plans, denoting a sense of urgency in producing the visions and in their applicability, as well as the importance of making them ongoing—and thus sustainable—through capacity building.

The availability of municipal funding and technical expertise in Medellin most likely surpassed what is available in most developing cities. However, the main takeaway here, in terms of capacity building for the tangible improvement of informal settlements and how these efforts can impact greater urban agglomerations, depends on the following factors:

- Political commitment to address an accumulated social debt of decades of exclusion, which, in the case of Medellín, was exacerbated by deep-rooted violence-related social unrest, guerrilla movements and drug trafficking. Authorities making informal settlement upgrading programs the main course of action and envisioning them as part of a wider urban strategy. In this case, the administration targeted the most challenged and violent peripheral neighbourhoods, demonstrating that rapid change was possible, even operating under the most adverse conditions.
- Efficiency, Proximity and Networking. Mayor Fajardo realized that it was crucial to assemble efficient, knowledgeable, and honest teams, capable of delivering high-quality design solutions, while providing sustained management to the programs. At a municipal level, the different branches of the administration would work as a tight team transversally (departments in charge of infrastructure, social

[4]Medellín won the 'Innovative City of the Year' award, organized in 2013 by The Wall Street Journal and City, as the most creative city of the year, from a shortlist that included New York City and Tel-Aviv. This was one of many prizes and praises that Medellín received, considered for decades the most violent city in the world. Medellín exemplifies that even under the most adverse conditions urban improvement is possible, and in a short time, becoming a reference for many developing nations and particularly in relation to the improvement of informal settlements. See Wall Street Journal. 03 14, 2013. http://online.wsj.com/ad/cityoftheyear (accessed 18 01, 2019).

housing, public spaces, community services, cultural development, health, etc.), holding weekly meetings with the mayor, and with the participation of project coordinators. The coordinators were usually young and energetic professionals who would be on site every day, and sometimes even resided in the communities, in order to establish communal networks and push ahead the programs. These on-site facilitators were asked to brief the municipal teams on the advances and to be candid about bottlenecks and limitations, forcing the different branches to accelerate their responses, operating under a unifying vision. In order to make them effective, it was clear that the programs had to be conceived and carried out on site, in close relation with the community leaders and proactive residents, gaining mutual trust, sharing information, responsibilities, and even working in tandem on planning, design, and execution phases. Holding participatory workshops in the settlements in relation to the ongoing work and to future actions became a normal practice, acting as a very powerful tool for capacity building—to the extent that, when the municipal administration changed hands to other political parties, the communities lobbied to continue with the programs as they were originally conceived.

- Delivering with a sense of urgency and the strategic staging of interventions. In Medellín, after decades of extreme violence and an isolation from the national, regional, and global markets, Fajardo and his talented and committed group of collaborators realized that they had a unique opportunity to make a difference, if they acted not only coherently but also promptly.

It is important to note that, in Latin America, traditional urban plans were usually mostly data oriented, costly, and time consuming, and they were often delivered after the municipal teams that initiated them had left office, focusing on entire metropolitan and urban systems. When these plans were finally enacted, the urban conditions had changed, and very few actions that could make a difference were actually delivered.

In Medellín, site-specific acupunctural visions, called Planes Urbanos Integrales (PUIs), or Holistic Urban Plans, were intended to induce immediate beneficial changes, gain the interest and support of the communities, which were tired of undelivered promises, and conduce to further and carefully orchestrated interventions at different scales.

The administration was committed to making a significant change in a very short period and to demonstrating that politicians could operate in a different way during the four years they would be in office. Roughly speaking, the programming of activities followed this agenda: one year to produce a solid site analysis of each neighbourhood chosen and to envision the district plan; one year to deliver the pilot projects and construct them; one year to start operating them, training personnel, carrying out community activities to ensure participation and engagement; and, finally, one year to evaluate results, change courses if required, and advance further projects in a carefully orchestrated manner.

The selection of the first areas to be tackled was pivotal. In order to have a larger impact, the administration analysed GIS information to determine which were the most challenging neighbourhoods—usually the ones where urban and social indicators of levels of violence, inaccessibility and isolation, lack of infrastructure, public spaces and communal services usually overlapped. This helped determine the areas that would be targeted first: the Comuna Nororiental, San Javier, La Ladera, and Moravia neighbourhoods. The first three neighbourhoods were located on very steep terrain, while Moravia was in a lower elevation partially built on the flood plain of the Medellin River, and above and adjacent to a now-defunct landfill.

The projects for the Comuna Nororiental and San Javier envisioned the introduction of Metrocables, aerial gondolas or cable cars (used in non-tropical countries as sky lifts), in order to provide quick transportation, overcoming the topographic differences—up to 300 m—and the physical barriers such as streams and cliffs. The construction of these gondolas would require minimum displacement of homes in order to build the pillars for the systems, to locate the stations in strategic points within the dense settlements, and to create public spaces and services in proximity to the stations. The construction of new substitution housing was an upfront requirement in order to provide these open spaces, to introduce communal services, and also to relocate residents from high-risk areas, such as the floodplains of creeks, built on unstable land, or over landfills. The new homes were constructed as formal dwellings in walk-up apartment buildings, located within the same neighbourhoods. Only in the case of Moravia, where there was no land available to do so, some residents had to be moved to the higher elevations of the San Javier neighbourhood three miles away, served by the new Metrocable, disrupting the existing socio-economic activities of the residents who were previously closely attached to this challenged site. In all cases, numerous community meetings were held in order to reach an agreement with the people who would be relocated, accepting their new dwelling arrangements, without which the holistic improvement plans could not have been delivered.

High-quality libraries or community centres were provided in all these neighbourhoods; their projects, resulting from open design competitions, were to be located in prominent locations adjacent to new public spaces, accessible to residents and visitors alike. These educational and leisure facilities resulted in symbols of the urban and social transformations that had been achieved. In San Javier and in La Ladera, the new libraries were erected on sites that were previously occupied by jailhouses, which had to be relocated, erasing the negative stigma brought by the penitentiaries that had been the origin of both neighbourhoods. Particular care was given to the treatment of public spaces as the framework that connected all these interventions, an opportunity to improve infrastructure and to create well-designed places for social interactions. The programming of activities and events for populations of all ages was also an important goal in order to bring life into these new public spaces and create a sense of belonging and attachment.

As these projects were being carried out during a period in which important legal, managerial and political changes were taking place in Colombia on a national level, which allowed the country to gradually regain governance after decades of prolonged drug violence and guerrilla warfare, the decisive municipal commitment and the holistic visions of the projects executed in Medellín under municipal leadership, together with the quality of the infrastructure, transportation systems, communal facilities and public spaces provided, clearly began to transform these challenged neighbourhoods into healthy communities. The significant reduction in the levels of violence, the return to these neighbourhoods of populations that had been forced to leave during periods in which many of them were considered war zones under the control of drug lords, the emergence of economic activities catering to locals and to visitors, the improvement and vertical growth of the self-constructed buildings, and a new sense of pride among residents in relation to all these changes are powerful indicators of the success of these initiatives.

While the PUIs completed for the most challenged informal neighbourhoods in Medellín were a strategic priority, plans were also conceived and actions delivered in other areas of the city with the same efficiency and high standards, including: in districts where formal and informal constructions coexisted, providing urban spaces, services, and amenities of a metropolitan nature; in the old urban core of the city; in important urban corridors formed by ravines, segmenting residential districts of different social strata, etc. One such project is the recreational and pedagogical district adjacent to the Moravia neighbourhood, which includes the refurbished botanical garden, the new Parque Explora, a mixture of science centre and children's and youth museum, and the retrofitted planetarium adjacent to the new Parque de Los Deseos, served by the Universidad metro station. All these metropolitan-scale amenities are interconnected and linked to the Moravia neighbourhood by the thoughtful landscape intervention along Avenida Carabobo, which was one of the historic access roads to the city. This road was previously a congested and deteriorated artery but was redesigned as a broad pedestrian-friendly open space, from which traffic is detoured particularly on weekends. Before these interventions, almost nobody visited the area. This district is today intensively used by residents of lower-income districts and by people who reside in other parts of the city, as well as by tourists.

The incremental and highly calibrated modality of interventions that began during this period is still embedded in the modus operandi and the urban culture that put Medellin on the map as global point of reference for urban regeneration and social inclusion. The Medellin experience has tried to be emulated in different Colombian cities and in other countries, perhaps not with the same level of success, probably because in many cases only some of the technical components have been replicated, not understanding the importance of the holistic community-based, and thus ongoing, approaches implemented in Medellín.

For instance, in Caracas and Rio de Janeiro, they have incorporated Metrocables in some of their most challenged neighbourhoods, using the same technology as in Medellín. In the San Agustín neighbourhood in Caracas, the stations are not appropriately located, they are surrounded by poorly designed open spaces, fenced off from the communities, the informal homes are being gradually demolished to be replaced by social housing projects that the original residents cannot afford, and so on.

In a recent interview[5] that I held with the architect Alejandro Echeverri for Professor Lautaro Ojeda, Director of CINVIT (Centre for Research on Informality

[5]https://www.youtube.com/watch?v=9En3Ig1uSOc&t=49s.

of the Universidad de Valparaiso, Chile), Echeverri posits that, despite the fact that there is worldwide awareness of the importance of addressing informality as an integral component of the cities of the Global South, it is surprising to see that the Medellín experiment has not been widely replicated, partially reproduced or improved in other contexts. He says Medellín should not still be such an important reference; there should be many and better examples by now. We both believe that the problem seems to be the lack of political will to acknowledge the dangers of inaction, to take on responsibilities and carry out these plans, which require very hard work, talent, communal engagement and transparency—attributes which do not usually fit in with the political agenda. In the same interview, I refer to the importance of such initiatives, but also to their limitations, and to the need to explore new paradigms to cope with the emergence of new informal settlements of gigantic proportions. These are the topics that are addressed to wrap up this article.

6 On Limitations of the Plans for the Rehabilitation of Informal Settlements

While these programs drastically reduced the levels of violence, boosted the self-esteem of the communities, reconnected them to the formal city, improved accessibility, relocated settlers from high-risk areas in substitution housing within the same districts, triggered economic activities, provided high-quality public spaces, infrastructure, communal services and amenities, and changed the perception of how inhabitants of the formal city consider informality, they also had limitations. Acting on highly consolidated settlements is a laborious process, in order to gain the acceptance, trust and engagement of the communities—particularly when a certain degree of relocation is required in order to free up space for the interventions described. Relocation requires negotiation skills to convince those who will be displaced that they are making a good deal. They are time consuming, technically challenging, and costly.

Additionally, it is very difficult to provide services, infrastructure, and public spaces such as large parks, technical schools, universities, food distribution and transportation centres, hospitals, large sport facilities, manufacturing and industrial areas, and so on, without disrupting the urban and social fabric. For instance, while programs to improve the self-constructed neighbourhoods of Medellín are highly praised by the residents and have served as academic and institutional references in other contexts, the majority of adult residents still have to commute to the formal city for access to better jobs and metropolitan services. These weaknesses are magnified when the informal areas become very large and at a greater distance from the formal areas. In other words, even the best plans for the rehabilitation of informal settlements have limitations.

7 On Capacity Building for New Informal Settlements

UN-Habitat and other institutions foresee that, in less than 25 years, people living in informal settlements will double today's estimates, reaching 2 billion inhabitants;[6] in other words, 1 billion more inhabitants will self-construct their habitats, occupying new territories. This represents a major environmental, socio-economic and political challenge if left to grow spontaneously.[7] Furthermore, surgical upgrading of these new settlements, once they are consolidated, will have limited effects, if any, on the wellbeing of these informal mega-conglomerates. Innovative planning, design, and managerial paradigms are urgently needed to foster the exponential growth of the self-constructed city in a pre-emptive manner, from the early stages of occupation through all the different stages of transformation. Planning ahead and acting in tandem with community efforts will allow for addressing the limitations of the surgical approach by selecting adequate sites, accompanied by a compelling land-banking policy, and envisioning and securing the spatial requirements of the public domain, including essential support systems such as infrastructure, mobility, public spaces, water management, urban agriculture, community services, economic drivers, etc.

Criteria and tools to orient these efforts in a pre-emptive and transformative manner are being advanced at an academic level by several institutions. My work[8] includes a general appreciation of the lessons learned from the programs to improve existing informal settlements, which have served to nurture this new approach to addressing the upcoming challenges and responses of new self-constructed settlements, as well as predominantly informal larger urban systems. There seems to be virtually no understanding of the implications of not addressing the future growth of the self-constructed city and even strong resistance to accepting the need for assisting in its development: i.e., planning, designing, inducing, and managing the dynamism of emergent informal settlements as the genesis of a greater metropolitan system. There is still the stigma of referring to them as illegal or irregular. Informal settlements are still seen as the problem—and not a vital component of the solution, if appropriately guided. The task of tackling entire new predominantly informal cities will require even stronger political commitment and technical skills due to the multiple variables, scale, and timespan involved.

---

[6]Ibid note 1.

[7]Ibid note 2.

[8]Available in two publications and an online lecture: D. Gouverneur, Planning and Design of New Informal Settlements-Shaping the Self-Constructed City, Routledge, Oxford, 2014; D. Gouverneur, Diseño de Nuevos Asentamientos Informales, Universidad de La Salle, Bogotá/Urban-EAFIT, Medellín, 2016; and https://www.coursera.org/learn/designing-cities/lecture/G5jjZ/rapid-urbanization-and-informal-settlements.

8 Some Ideas to Assist New Informal Settlements, Fundamentals of the Informal Armature Approach (IA)

Since the emergence of the Site and Services programs in the 1960s, not much has been done in terms of thinking ahead and acting on the growth of self-constructed settlements. We are in urgent need of new paradigms to foster new informal urban growth. We must implement methods, utilizing tools capable of tapping into the potentials of informality and engaging with the communities that self-construct their habitats. This should be done keeping in mind that, in many cases, we are seeding very large and complex urban areas. We should draw upon innovative design and performative ideas. Our aim should be to steer the predominantly informal cities toward higher levels of performance and help their residents attain decent living conditions (Fig. 1).

How is the Informal Armature (IA) approach different from previous methods? The IA is intended to be simple in its principles, design ideas, and applicability. It aims for compelling results. It suggests a symbiosis between the formal and the informal modes of city making, which can result in rich and dynamic urban ecosystems that are highly competitive and resilient. The key conditions of the IA approach are as follows: pre-emptive and transformative; physical and performative; hybrid and multi-scale operations; environmentally friendly; and simple. These conditions are all closely interrelated.

IA is based on flexible design components, adaptable to different site conditions.

Let us delve into these conditions:

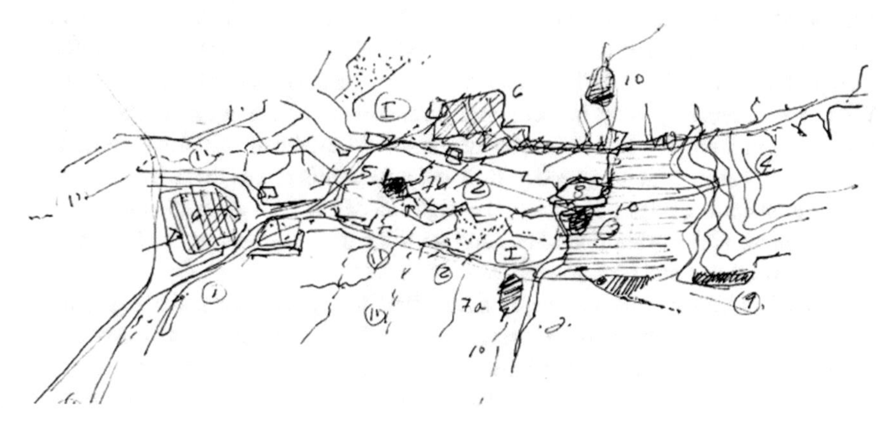

Fig. 1 Sketch of an informal armature by David Gouverneur

8.1 Pre-emptive and Transformative

The Informal Armature approach is essentially a pre-emptive planning and design method that aims to foster the growth of the predominantly informal city. It intends to serve as a support system, enhancing the internal dynamism of informal urbanization. The main premise is that planning ahead and guiding the urbanization process allows for more balanced and efficient development. This contrasts with retrofitting consolidated informal neighbourhoods. In order for the IA to be effective, the following conditions must be met: (a) political acceptance that unassisted informality will occur on its own, occupying inadequate locations without the basic urban frameworks, and thus will result in difficult urban situations with severe social and environmental implications; (b) the availability of suitable public land to apply the approach, the will and capability of the public sector to assemble it, acquire it, or develop it in partnership with the private sector; and (c) the organization of teams of facilitators, committed to the IA initiative. The teams must also have the ability to adapt the IA criteria to local conditions, engage the community, incorporate expertise according to particular needs, and foster the transformation of the IA sites until their assistance is no longer necessary.

8.2 Physical and Performative

Traditional planning in most developing countries is reduced to static land-use plans with no design considerations; they are centred on controlling the real estate market, which has no incidence on informal urbanism. Developing countries frequently blame their problems, and the inability to cope with them, on the lack of financial resources. While there certainly may be economic limitations, greater weaknesses in these nations stem from the lack of coherent visions to establish priorities, coordinate actions, and ensure efficiency and transparency. Medellín points to the importance of introducing cutting-edge planning and design standards, with efficient and sustained management, and working hand-in-hand with the communities as the best way to ensure deliverance and accountability.

Thus, the IA approach combines notions of spatial organization and design criteria with on-site management. It suggests how to foresee simple spatial organization patterns to attract and guide the initial occupation of the sites and create a strong bond with the settlers. The strength of the approach relies greatly on establishing trust between the facilitators of the program and the communities, and also on how urban changes are managed. The transformations of the IA territories will not only occur in the public realm, but also through the appropriately timed inclusion of a diversity of uses that are meant supplement the self-constructed areas, as the communities consolidate, evolve, have greater expectations, and become part of much larger urban agglomerations.

One of the main lessons gained from the Site and Services programs and the informal settlement improvement plans is that the spatial organization and physicality of the self-constructed city are important. Sustainable urbanism requires quality planning and design, and this may include: designing compelling and performative urban landscapes; facilitating entrepreneurship; diminishing energy consumption and waste; enhancing social interaction; reducing violence, favouring shared participatory governance, and so on. Facilitators of the IA should have the ability to make the early settlers feel that they are a proactive part of improving their districts and help them engage in all facets of the foundation and evolution of the settlements.

8.3 Hybrid and Multi-scalar Operations

Both Site and Services and informal settlements improvement projects seek to merge the dynamics of the self-constructed habitat with planning and design. IA operates with the same logic, but its goal is to attain a highly balanced and competitive urban product that will, over time, become the mainstream, as the dominant form of urbanization in many developing countries will be informal settlements. In order to do so, IA facilitators must develop the skills to deal with early phases of occupation, deploying rather simple design components in accordance with the basic needs of the new settlers, while simultaneously planning and preparing to address the urban and metropolitan demands of larger urban ensembles. This is not an easy task, since it requires skilled "navigators".

Tracts of land for self-constructed dwellings that can be subject to different types of urban frameworks and technical assistance, as envisioned by the Sites and Service programs—accompanied by other forms of social housing—are very important components of the IA fostered territories, but far from being the only ones. The IA initiative considers strands and patches of vacant land that may initially incorporate simple mobility systems or allow for the installation of reception tents with a provisional supply of potable water. These open spaces may eventually sustain an efficient public transportation system and a diversity of uses, civic amenities, and compelling public spaces. The IA program may also predefine how to protect waterways and areas of vegetation that will become the recreational corridors for the city or envision sites employed initially as recycling centres in order to obtain construction materials to assist the settlers in building their shelters. These might be areas that can eventually serve to initiate urban agricultural programs which, in time, will give way to manufacturing centres, technical schools, large city markets, and so on.

Some land, with temporary uses, may be kept in public hands until the aspirations and income levels of the community demands rise. The private sector can then provide more sophisticated commercial areas and amenities. These operations may represent opportunities for the IA facilitators to attract private investment, targeting mixed-use developments for higher income groups, providing formal housing, offices, cinemas or recreational facilities, which the original settlers will also demand. The additional value obtained from such operations down the line will help the IA facilitators generate revenue to reinvest in other services within the same territories, or to reinitiate similar IA initiatives on new urban frontiers. These IA-managed territories are expected to thrive with the vitality and character of the self-constructed habitat, combined with the amenities of the formal city, as a true hybrid product.

8.4 Environmentally Friendly

Responsive contemporary urbanism has the obligation to address environmental aspects that are pivotal not only for the wellbeing of communities but also for the health of the planet. Such considerations have not been on the agendas of conventional urban planning, which still determines the urban conditions of the formal city in many developing countries, nor were they considered in the Site and Services programs as precursors of the IA approach.

Since IA begins operating before occupation takes place and determines the first physical and performative conditions of the settlements, it represents a favourable and still flexible milieu to introduce environmentally friendly criteria. Among the some of the most sensitive aspects we can mention are: mechanisms to protect adjacent watersheds; areas of biodiversity; valuable agricultural soils; or the hydrological network, adjacent to and within the urban areas. Similarly, IA can facilitate the introduction of green infrastructures, pedestrian and clean public transportations networks, recycling programs, community agricultural gardens, etc. The presence of proactive IA facilitators will allow for the planning, design, and management of the new informal/formal districts with these ideas in mind, making them meaningful for the early settlers as part of their daily lives. It is important to keep in mind that, in non-fostered informal settlements, particularly in the primary phases of occupation when they are in survival mode, settlers may not be aware of or care about environmental issues. Many of them arrive from rural areas and quickly lose their agrarian skills and familiarity with nature as they become urbanites. The IA allows for accommodating these cultural assets in a new urban setting.

8.5 Simple: Based on Flexible Design Components, Adaptable to Different Site Conditions

Throughout history, cities have been experimental in testing new urban organizations. Some have been shaped as prototypes embedded in cultural practices, and others have been introduced by leaders, designers, economic interest groups, providers of services, etc. Design solutions are often replicated in contexts different from the ones in which they originated. Those that prevail tend to be simple in terms of how they are spatialized and implemented. Planning and design prototypes used in the developing world often respond to particular ideologies and political, economic, and social goals but do not always benefit the majority of the population. IA presents a toolkit of simple design components whose main goal is to provide better living conditions for the largest segment of the population in the developing world, the poor. These components are not dogmatic in their nature, scale, or in their morphological and performative qualities. Rather, they embed principles that can sustain the multiple processes mentioned in the preceding sections. Their successful deployment as spatial and performative solutions will depend on the local contextual and cultural nuances and on the ability of the IA facilitators to engage the communities and other urban actors in the initiatives.

To simplify its application, the IA approach groups the design components into the following categories: Corridors, Patches, and Custodians. These components are expected to act as urban organizers and as an integrated system, supporting the informal initiatives and performing different roles.

The Corridors are essentially linear components or strands that structure the public domain, acting as the skeleton and DNA of the systems of open spaces, infrastructure, mobility systems, water management devices, services, and so on. The Corridors are subdivided into two sub-categories:

- Attractors: act as pullers or hot spots, encouraging the concentration of the population and activities towards areas where it is more appropriate to do so, providing spatial and managerial conditions to better handle them, and
- Protectors: contrarily, dissipate urban energy, safeguarding environmental assets and acting as buffers to avoid urban occupation or expansion—whether informal or formal—in sensitive areas (Fig. 2).

Attractors and Protectors are expected to manage a high percentage of the transformations of the settlements into larger districts and urban territories, and they require a higher degree of public involvement to achieve their goals, particularly in the early stages of occupation, consolidation, and growth.

Fig. 2 Composite diagram of attractors and protectors by David Gouverneur. An example of urban scale corridors for Chitungwiza, Harare. Zimbabwe, by Megan Talarowski and Peter Barnard. Advisor: Gouverneur and Thabo Lenneiye

Fig. 3 Left: Composite diagram of Patches by David Gouverneur. Right: Proposal of productive patches (greenhouses) with protected wetlands for La Sabana de Bogotá, Colombia. By Tamara Henry. Advisors: David Gouverneur and Abdallah Tabet

Patches represent the areas of urban infill that are nurtured or held together by the Corridors. They are also grouped into two sub-categories:

- Receptors: sites made available for informal occupation, where the self-constructed neighbourhoods are expected to flourish; and
- Transformers: areas capable of sustaining a multiplicity of uses which, together with the Corridors, are expected to differentiate the IA from the non-assisted informal settlements and from the Site and Services programs (Fig. 3).

And, finally, the Custodians are components that may appear in all the previous categories to ensure the availability of space needed to gradually accommodate the urban demands, as the IA territories evolve (Fig. 4).

One of the difficult tasks for the success of the IA approach is securing the spatial requirements to gradually introduce more complex urban uses, services, and infrastructure at more mature stages of consolidation of the settlements, according to the economies of scale associated with much larger populations and urban areas. If these spaces are not secured, the mega-informal cities of the future will result in a quilt of predominantly residential patches with some local commerce and services. They will continue to suffer from strangled mobility, limited productive jobs, in addition to a lack of metropolitan services and amenities, and large open spaces. The current living conditions and the diminished performance of Mexico City, Lima or Caracas provide a glimpse into the future of the unassisted, incomplete, and marginal mega-informal cities.

The Informal Armatures approach, at the core of the book Planning and Design for Future Informal Settlements: Shaping the Self-Constructed City, is a call for urgent action, directed at politicians, academics, professionals, institutions, the private sector and communities alike, to embark in a simple but committed way in stewarding the informal city that is just emerging today, seeding a sustainable future. With this

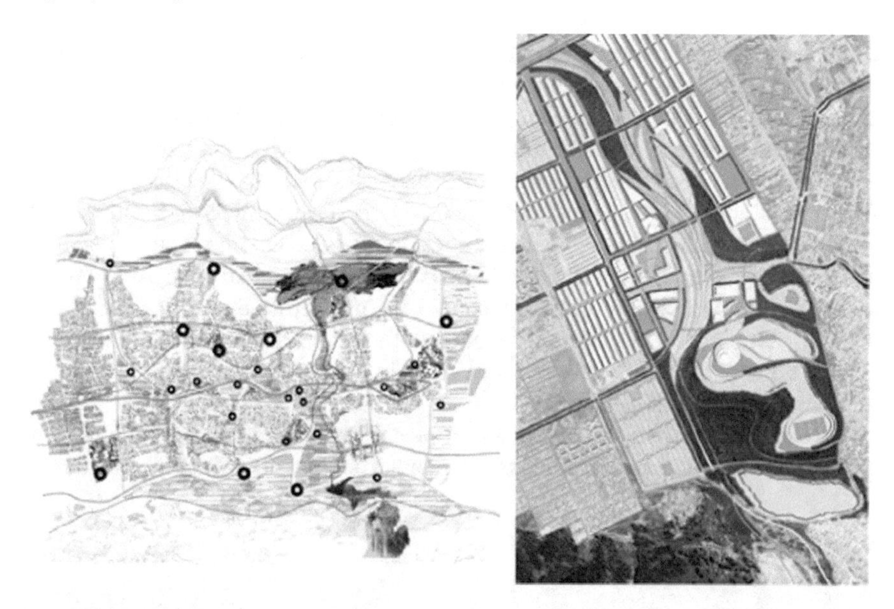

Fig. 4 Left: Composite diagram of custodians over corridors and patches by David Gouverneur. Right: Proposal for cultural and recreational facilities as Custodians of an extinct quarry, in Bogotá, Colombia, by Rachel Ahern. Advisors: David Gouverneur and Abdallah Tabet

Fig. 5 Academic proposals testing the Informal Armatures Approach by students of PennDesign from the University of Pennsylvania. Example of armatures for informal/formal expansion in Chitungwiza, Harare, Zimbabwe, by Daniel Saenz. Advisors: David Gouverneur and Thabo Lenneiye

chapter and book, I sincerely hope that it will attract the attention of those who have the responsibility, the means, and the ability to induce urban changes in developing cities and motivate them to implement pilot projects to prove its validity. The book includes academic examples illustrating the physicality and performative conditions that the approach may handle, in very different contexts. It also suggests avenues for further experimentation and research, which the IA requires (Figs. 5, 6 and 7).

The lives of billions of people in the developing world are impacted by planning and design paradigms that were introduced decades ago. Similarly, the lives of many more will be adversely affected by inaction. IA aims to prove that it is a feasible and appropriate approach to make a difference. Informal settlements are continuing to emerge without support in thousands of cities in the developing world. Now is the time for urgent and efficient action.

We can conclude that one area in which capacity building is needed the most is in creating widespread awareness and in training the people who will be the facilitators

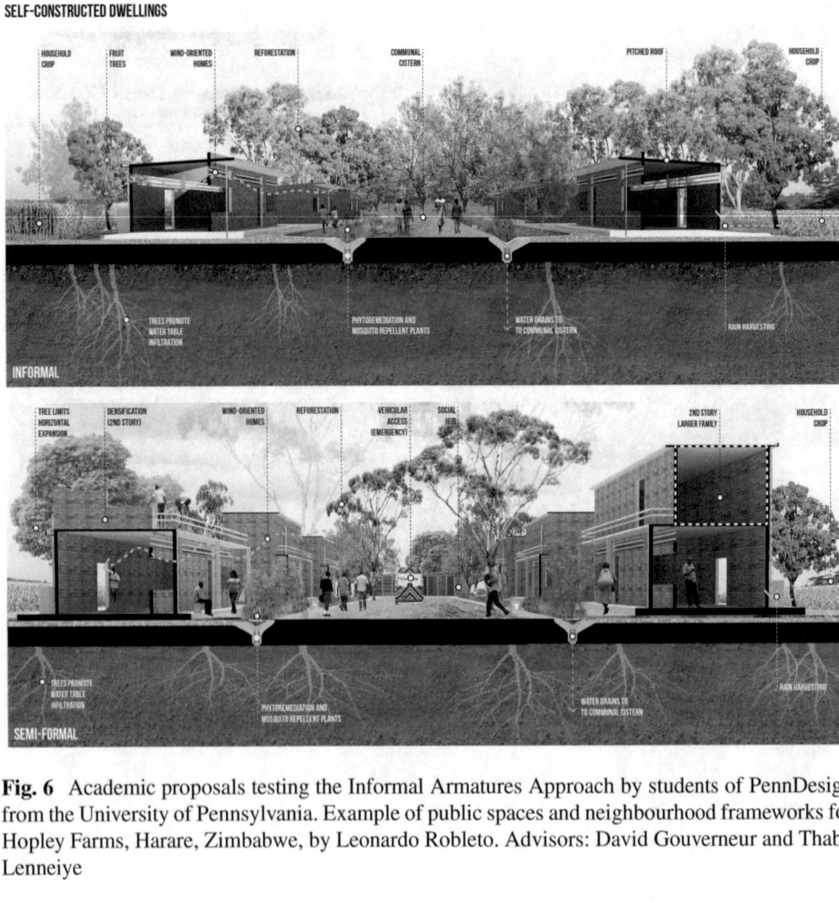

Fig. 6 Academic proposals testing the Informal Armatures Approach by students of PennDesign from the University of Pennsylvania. Example of public spaces and neighbourhood frameworks for Hopley Farms, Harare, Zimbabwe, by Leonardo Robleto. Advisors: David Gouverneur and Thabo Lenneiye

Fig. 7 Academic proposals testing the Informal Armatures Approach by students of PennDesign from the University of Pennsylvania. Armatures to relocate settlers currently on unstable land, Barrio Santo Domingo, Medellín, Colombia, by Kimberly Cooper, Rebecca Fuchs and Keya Kunte. Advisors: David Gouverneur and Trevor Lee

of this process. As in the case of Medellín, and many other milestones in city development, innovation usually emerges within academia, or in the minds of professionals who have had exposure to cutting-edge ideas and hands-on experiences. However, the Medellín success story also demonstrates that political support is perhaps the main component in the equation, and that advancing pilot projects, working with the best technical teams, engaging communities with vigour, advancing programs with calibrated actions, and delivering with high-quality planning, design and managerial standards, may the best scenario for capacity building.

References

Baldó J (2007) El programa de habilitación de barrios en Venezuela: Ejemplo del control del proceso de construcción y de administración de los recursos por parte de comunidades organizadas. Tecnología y Construcción [online] 23(1):9–16

Gouverneur D (2006, 03 XX) De los superbloques a los asentamientos informales. Concepciones disímiles, resultados similares. Retrieved 07 29, 2013, from La ciudad viva. http://www.laciudadviva.org/opencms/export/sites/laciudadviva/recursos/documentos/De_los_Superbloques_a_los_Asentamientos_Informales.pdf-ee21e2583c667528b8c78f69be3970e6.pdf

Gouverneur D (2009) New forms of urbanization. Cuadernos Unimetanos. Prof. Flores., Caracas

Gouverneur D (2014) Planning and design for future informal settlements: shaping the self-constructed city. Routledge, Oxford

Grauer O (1991) Principles, rules and urban form: the case of Venezuela. University Microfilms International
Grauer O (2002) Democracy and the city. In: Democracy in Latin America. David Rockefeller Center for Latin American Studies, Harvard University, ReVista
Potts D (2010) Circular migration in Zimbabwe & contemporary Sub-Saharan Africa. Boydell & Brewer Inc., Rochester
United Nations Habitat (2003) The challenge of slums. Global report on human settlements. Official, Earthscan Publications Ltd., London
Wall Street Journal (2013-14-03) Wall Street J. Retrieved 2013-02-08 from http://online.wsj.com/ad/cityoftheyear

Balancing Participatory Design Approaches in Slum Upgradation: When Top-Down Meets Bottom-Up!

Reena Tiwari, Jessica Winters, and Neeti Trivedi

Abstract The chapter discusses a participatory planning approach utilised for informal (slum) settlement upgradation. A review of approaches identifies the failures of top-down processes and demonstrates the challenges of bottom-up processes. A middle-ground planning approach, where top-down spatial solutions grounded in disciplinary knowledge mesh with the bottom-up spatial, economic, social and ergonomic constraints of slum dwellers, is proposed. When the appropriate balance is achieved, design creativity and inclusive participation result in greater capacity for the slum dwellers.

Keywords Informal settlements · Participatory planning · Slum upgrading · Capacity building · India

1 Introduction

The one-size-fits-all (Albrechts 2013) approach to informal or slum settlement upgrading with the traditional tools of master plans, development project reports, biometrics, spatial development plans, implementation plans, investment plans, etc. has frequently served the needs of governments and other stakeholders instead of the slum dwellers. This has resulted in a need to promote community-led, participatory planning to improve these settlements. Formalisation of these communities, a result of the top-down only approach, and failure to implement programs strategically and in a timely manner, a result of the bottom-up only approach, has shown that a balance needs to be negotiated between the top-down and bottom-up approaches. Bottom-up participatory approaches have also been criticised as tokenistic, simplistic and inflexible, resulting in inauthentic and ineffective outcomes. Analysis of the top-down,

R. Tiwari (✉) · J. Winters
School of Design and Built Environment, Curtin University, Perth, Australia
e-mail: R.Tiwari@curtin.edu.au

N. Trivedi
BN College of Architecture, Pune, India

O. Carracedo García-Villalba (ed.), *Resilient Urban Regeneration in Informal Settlements in the Tropics*, Advances in 21st Century Human Settlements,
https://doi.org/10.1007/978-981-13-7307-7_7

bottom-up paradigms through literature review and case study analysis indicates that a hybrid, middle ground approach is preferred. The authors review three projects that incorporate a participatory approach and are internationally recognised as examples of successful physical upgradation for the urban poor.

Yerwada Slum Upgrading Project (YSUP) in Pune, India is a large-scale collaborative project where power dynamics between stakeholders significantly affected the project outcomes. The Innovation Centre for Poor (ICP) in Ahmedabad, India is a small-scale project involving a single institution. The aim was to improve working and living conditions of households engaged in home-based industries in a slum settlement. The last case study is the Ramdev Pir Tekro Activity Centre project, a medium-scale project in Ahmedabad. Here, we deconstruct a collaborative and ethnographic model of engagement involving residents of a large slum community and multiple stakeholders.

2 Background

The 2009 Global Report on Human Settlement 'Planning Sustainable Cities' identified informality as one of the factors shaping 21st-century cities. Informality has been exacerbated by problems in four areas: employment; social development; urban growth; and fulfilment of basic needs. The traditional urban planning approach of 'formalising' informal settlements usually fails to incorporate informal settlements into the formal city or to address the social, economic, cultural, and urban challenges that they face (Buehn and Schneider 2012; Elgin and Oztunali 2012).

Most urbanising planning practices lack a critical analysis of 'what', 'where', and 'who', nor do they produce a robust and tangible approach to deal with the problems of urban areas. The World Bank has promoted community participation as an approach in which the 'who' is defined as the 'local community', and that community is then responsible for defining 'what' and 'where'. The objective was for redevelopment projects to 'reach the poorest and the vulnerable in the most effective and economical way, by sharing costs as well as benefits, and through the promotion of self-help' (Craig and Mayo 1995, 2). These programmes came to be known more for cost saving than for political and social risk-taking. This led to further decline in the social and economic structures of the urban poor, while appearing to promote the importance of 'community' (Craig 2007). In reality, the development philosophy that emerged, shaped largely by national and international organisations, gave inadequate consideration to the problems of social justice and disregarded the identity and self-esteem of the poorest and their right to contribute to the decision-making processes that concern them.

A further challenge lay in defining 'community', although it was recognised that the community/public interest was fragmented, and that a new pluralism must engage with a broader range of interests to manage the environment. In 1990, Healey predicted that citizens would increasingly demand more say in how their environment is managed and would expect accountability from all agencies, through democratic processes. Substantial conflict regarding economic, social and environmental goals would emerge from a more pluralist society. Resolving these conflicts in democratic ways would be critical (Healey 1997). The new pluralism requires planning processes which facilitate dialogue between different actors to address and manage conflict, and which enable people to express their concerns, needs and ideas. Various techniques of community participation have been used to draw these informal marginalised communities into the mainstream urban communities. The next section reviews some of these in the Indian context.

3 A Review of Participatory Approaches

In response to the need for more engaged communities in development projects, participatory methods have 'emerged and evolved over time and space' (Mubita et al. 2017) and become a core component of mainstream development thinking and practice (Parfitt 2004). Participation is essential for community-led planning to deliver 'genuine and sustainable poverty alleviation' (Nuttavuthisit et al. 2014, 56) by incorporating local knowledge into development practices (Mubita et al. 2017; Gegeo 1998), forming a 'bottom-up' approach and allowing local people to define their own needs and actions (Nuttavuthisit et al. 2014).

According to Parfitt (2004), the major difference between participatory approaches is whether the participation is seen as a means or an end in the development project or programme. Simply using participatory practices does not always guarantee poverty alleviation, empowerment of the marginalised, or successful capacity building (Mubita et al. 2017; Parfitt 2004). The adopted approach, as a means or an end, can significantly influence the success or failure of a project or programme.

Much participatory planning in major development projects has been criticised as tokenistic, rubber-stamping, a box-ticking exercise (Mubita et al. 2017; Parfitt 2004) or a means of saving money or avoiding political or social risk taking (Craig and Porter 1997). Although they may be called 'bottom-up', many participatory programmes are implemented in a way that is consistent with a top-down approach. Projects which include participation but are managed by 'traditional authorities' and guide participants toward predetermined project goals and objectives (Parfitt 2004) are essentially a top-down operation, with participation 'evaporate[ing] once the task is completed' (Parfitt 2004, 539).

The language used in academia to support or justify the use of participation is loaded with terms used in the top-down agenda. Examples include: effective, efficiency, sustainability, successful project delivery, validity, less costly, more timely, useful (Mubita et al. 2017; Parfitt 2004; Chambers 1994). Participation is driven by the need for the project to be successfully completed according to the business model of the development agency, 'Thus, power re-enters the equation incognito under the guise of demands of efficiency' (Parfitt 2004, 544). Sherry Arnstein's ladder of citizen participation provides an eight-staged framework to help provide a

Table 1 Ladder of citizen participation

1	Citizen control	Citizen power
2	Delegated power	
3	Partnership	
4	Placation	Tokenism
5	Consultation	
6	Informing	
7	Therapy	Non-participation
8	Manipulation	

An eight-staged framework of how power is distributed within participation
Author: Adapted and Modified from Arnstein 1969

clear understanding of how power is distributed within participation (Mubita et al. 2017; Arnstein 1969). The ladder has eight rungs with the lowest rung on the ladder being 'manipulation' and the highest being citizen's control (Mubita et al. 2017).

The eight 'rungs' are shown in Table 1.

When participation is used within agency frameworks, it becomes a tool to justify certain outcomes in bottom-up approaches. With limited community empowerment (Mathie and Cunningham 2003), participation remains on a lower rung of Arnstein's ladder of citizen participation (Mubita et al. 2017).

Some of the most documented bottom-up participatory approaches in the Indian context include Participatory Rural Appraisal (PRA), Appreciative Inquiry (AI) and Asset Based Community Development (ABCD). Although not all are relevant to slum upgradation (some have a more rural focus), it is important to highlight these approaches as they demonstrate the mainstream development practice. Many NGOs in India have adopted PRA to help 'express, enhance, share and analyse their knowledge of life and conditions, to plan and to act' (Chambers 1994, 1253). PRA ensures that the information shared by locals has high validity and reliability. Methods used in the programme include participatory mapping and modelling, transect walks, matrix scoring, well-being grouping and ranking, seasonal calendars, institutional diagramming, trend and change analysis, and analytical diagramming (Mubita et al. 2017; Chambers 1994). While the importance of the issues that PRA attempts to address (such as civil participation, social accountability and empowerment) are acknowledged, the approach does not go far enough to promote local involvement in financial management and implementation (Cornwall 2011, 271).

AI is also common in India and is based on the approach that emerged as a useful step in organizational change (Cooperrider and Srivastva 1987). Used frequently by Myrada (an Indian citizen empowerment group), AI focuses on two questions: 'What Is?' and 'What could be?' The system grounded in social constructivist theory has four stages—Discovery, Dream, Design and Delivery—based on a flexible and holistic programme focused on citizen empowerment. The guide prepared by Myrada and IIED (International Institute for Economic Development) (Ashford and Patkar

2001) assists in developing and implementing programmes in rural Indian communities. However, the competencies required of the AI facilitator/consultant and the training involved are specific, and not much has been written about this (Bushe 2011).

ABCD, a method inspired by AI, aims to allow communities to identify and mobilise existing assets 'thereby responding and creating local economic opportunity' (Mathie and Cunningham 2003, 474). Like PRA and AI, ABCD uses story collection, community asset mapping, steering groups, relationship building, leveraging resources from outside the community, and convening a representative planning group (Mathie and Cunningham 2003).

Critiques of PRA, such as Cornwall and Pratt's (2011), note that while the programmes can be fun and interactive, there are broader issues such as civil participation, social accountability, and empowerment to consider. This critique concludes that the objectives of financial management can override PRA objectives and that participation models are not 'simply about generating knowledge and information, but [are] complex sequences of engagement in different institutional interfaces and spaces' (Cornwall and Pratt 2011, 271). PRA's failure to accommodate external factors, particularly administrative and bureaucratic factors (Mubita et al. 2017), and the limited emphasis on power relations and structures (Parfitt 2004) can restrict its usefulness.

Although capacity building is at the core of the practice, ABCD still relies on creating core community steering groups, using interviews and storytelling as inspired by Appreciative Inquiry methods (Mathie and Cunningham 2003). The practitioner using participatory design must be flexible and adaptive to local needs and aware of local political and social influences. A participatory programme may be genuine and authentic but still ineffectual if it is not responsive to the context and flexible in its delivery. For example, a lack of response to the local political context might lead to an incorrect formation of community steering groups, which can be detrimental to authentic participation. Overly structured interviews and inappropriate questionnaires have proven unsuccessful in adapting to social- political dynamics (Chambers 1994).

Although participation is criticised as tokenistic in many approaches, any means of participation (even the weakest) can benefit those in developing countries. As Smith notes in Participation without Power (Smith 1998), participation is essential—although it would be naive to assume that all forms of participation are beneficial. If participation increases capacity and/or resources in a community, it can alleviate issues associated with poverty. One of the most effective outcomes of participation is to enhance existing community structures and to promote self-actualisation through capacity building. If participation is seen as an end (a product of the programme), it can enable the community to discover and develop independently (or in partnership) with an agency.

4 Case Studies: Slum Upgradation in India

4.1 Yerwada Slum Upgrading Project (YSUP) in Pune, India

Yerwada is a major slum settlement in Pune, India, with 22 slum pockets (PMC 2013; MASHAL 2011). There is in situ upgrading work in 12 communities. The case study discussed here focuses on Nagpur Chawl, which is the largest community with the highest number of houses due for upgrading. The upgradation project was an initiative launched by India's central government, under the Jawaharlal Nehru National Urban Renewal Mission (JNNURM), in 2005. Basic Services for the Urban Poor (BSUP), to be managed by the Ministry of Housing and Urban Poverty Alleviation (MoHUPA) (Patel 2013), were introduced as a component of the initiative.

Local NGOs with a strong community presence, previous experience of working with poor urban communities, and shared rapport with locally elected representatives were selected to be conduits between PMC and the community. As the project started developing, the NGOs collaborated with the local member of the PMC to set up large community meetings in the settlements to explain the project, its advantages, its timeline, parameters, subsidies and the community's contribution requirements, as well as individual households' financial capacity to arrange transit accommodation during in situ housing upgrading. The NGOs also appointed contractors and supervisors to do the ground work. The interface between the NGO and the community was limited to the meetings and was more mechanical. Instead, it was with the contractors and supervisors that the community had an opportunity to develop relationships. However, the contractors and supervisors did not necessarily have the skills or the time to involve community members. For them, in situ slum upgrading primarily meant demolition of the existing structures and construction of a contractor-built, government-funded house. The survey conducted by Trivedi in 2015 confirmed this when many households conveyed their dissatisfaction with the work of contractors and claimed that they had not seen the plan of the new house until after the demolition of their existing house. Many others asserted that the meetings were only to inform the community about the project, not to listen to community needs or concerns. The community was not an integral part of the project planning and implementation or the decision-making process, but merely a beneficiary. The criterion of building capacity of the community was assessed by Trivedi using a number of measures. Developing skills and building strong community networks as outcomes were rated 2/5, while involvement in decision making through programme management and problem assessment were rated 1/5 by the community residents who responded to the interviews (Trivedi 2017). The role of the NGOs should have been to make the voices of the poor heard in the project design and implementation. Instead, the reality was that, under government funding the NGOs were operating target-oriented services directly, by turning into implementers or contractors for the government development policy, rather than representing the local community of Yerwada.

Interesting power relationships impacted the outcomes of the project. In order to decentralise power and promote community participation, JNNURM—BSUP requires a transfer of power from the state to the local government understanding that sustainable improvements can only be achieved through social mobilisation, community participation, and collaboration between the local government agencies and NGOs. To achieve this, a tripartite partnership agreement was signed between the local government body—Pune Municipal Corporation (PMC), NGOs, and the community (individual households), wherein PMC was supposed to be the implementer, and the NGO would act as a link between the PMC and the community. The intent was to create a governance structure that would sustain and develop through decentralisation, thus bringing the local government and the community closer. Instead, the NGOs became the primary implementers on the project and retained contractors for the construction work, while working with a local leader to build community consensus for the project. Although the local leader had no formal role in the implementation of the project, he had strong connections with the community and was instrumental in bringing community members to meetings. Since he recognised the project as an opportunity for political advantage in advocating the benefits of BSUP to the community, his contribution was both positive and negative. While he helped the NGOs develop consensus, some political bias and favouritism cannot be ruled out. 28.2% of respondents during the survey confirmed this effect.

Although decentralisation and power-sharing were attempted in YSUP, there were complex political, technical and administrative challenges that were not taken into consideration. While adopting decentralisation and diversifying the sources of service delivery offers many advantages, it is important to recognise that central, state, and local governments may agree on project goals but perhaps have different priorities and strategies—which requires a strong institutional management capacity to guide the process forward.

4.2 Innovation Centre for Poor Project (ICPP) in Ahmedabad, India

The focus of the ICPP project was to understand and improve the working and living conditions of households with home-based industries. The aims of the project were 'to empower the urban poor by providing innovative solutions to improve earnings, enhance quality of life', and ensure long-term sustainable development through the mobilisation of the local community and local resources (Survey 2014). The project was initiated in 2009 and involved a collaborative effort between the following NGOs: MHT (Mahila Housing Trust), SEWA (Self-Employed Women's Association) Bank, SELCO (Solar Electric Light Company), and Foot Prints E.A.R.T.H (architectural consultancy in India). Also involved were academic researchers, architecture students, micro energy auditors, SEWA representatives, and skilled labour (fabricators/manufacturers and installers). The participants visited three communities to identify issues that women were facing in home-based work and to provide them with simple, case-specific solutions appropriate to the houses' conditions. A survey canvassed the space optimisation issues for home-based workers in three different occupations (Fig. 1) (Trivedi and Tiwari 2012). From an analysis of the data collected, the main issues identified were infrastructure, light and ventilation, shortage of space, and consequent health hazards.

Actively participating with residents in the dialogue and reflecting on the documentation gave the team a holistic understanding of the environment and conditions of the working women. The process was effective in raising awareness among resident

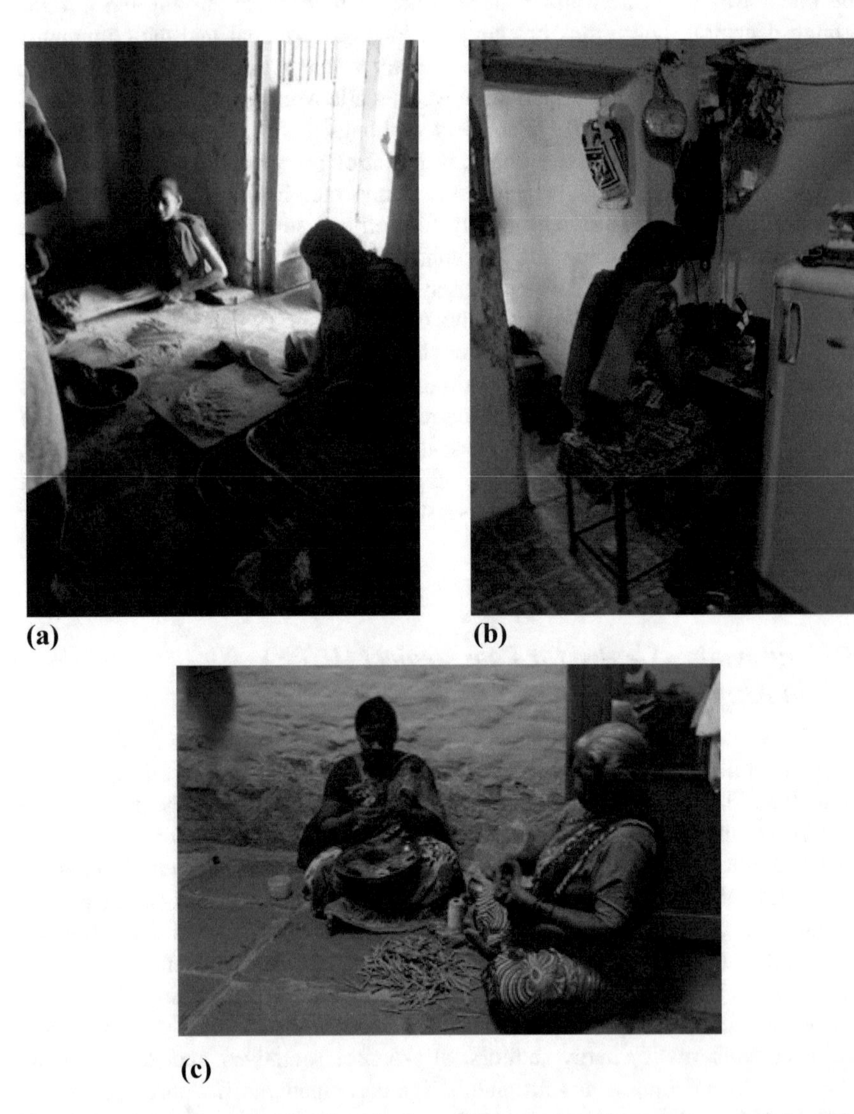

Fig. 1 Home based workers involved in three different occupations. **a** Incense stick making, **b** tailoring work, and **c** bidi making. Photos taken by Trivedi

women about space and ergonomic and health considerations critical to improving working conditions in home-based industries. The team generated solutions suitable for the women engaged in home-based work. These solutions reflected the lived experience of the voices and views of the poor in slum settlements (Nanavaty 2007). With a little guidance, solutions could be easily understood and implemented by the slum dwellers themselves. Case-specific strategies for individual homes were designed and varied according to professional demands. Products were designed to be functional and serviceable and to accommodate the requirements of the work (Trivedi and Tiwari 2012).

Based on data collected, the need to introduce natural light into dark spaces was identified (Fig. 2). Taking into consideration the housing conditions, the most suitable option for light and ventilation was a skylight called an Ujasiyu (Fig. 3). Installing the module on the roofs was outsourced to a fabricator and a welder. The module, associated machinery, ladder, and the welder were initially transported to the site by auto-rickshaw. The driver helped the welder and eventually learned how to install the modules. Given the circumstances, it was more economical to pay the auto-rickshaw driver to transport and install the modules rather than involve two separate workers. Thereafter, the NGO hired the driver for almost all the installations.

Using skylights has helped poor urban households reduce their electricity bills by up to 25%. The skylights also benefit home-based industries. Women can work

Fig. 2 Lack of natural light. Photos taken by Trivedi (2010, 2014)

Fig. 3 Installed Ujasiyu on the rooftops of the houses. Photos taken by Trivedi

inside while monitoring toddlers and children's schoolwork. Being protected from the rain, they can continue to work regardless of the weather, thus increasing their income. With the direct entry of natural light into the homes, output/production can be compared to that before the installation of Ujasiyu. Indoor air quality has also improved. Women making incense-sticks suffered from respiratory problems, since the powder in the air could not escape. After the installation of skylights, the powder can flow through the wire mesh in the roof, resulting in improved indoor air quality and fewer health issues.

Around 60 modules were sold in a year with the help of micro energy auditors, who—besides auditing the household energy usage and raising awareness of the importance of saving energy—acted as marketing representatives. Another 60 householders were exposed to the modules at a relative's or neighbour's house. The modules were sold for INR 2650 (US$40), including INR 1000 (US$15) in subsidy from the SEWA bank. Total investment per household was only INR 1650 (approximate, US$25). This could be recovered in a few months through savings on electricity bills. A significant decrease in demand for the installations occurred after subsidies ended and the price rose from INR 2650 (US$40) to INR 4500 (US$67.20). The project is however ongoing, and more purchasers of the product are being sought.

The ICP project has been successful in understanding the lifestyle, needs, and working conditions of the urban poor in informal settlements. However, the NGOs involved in the project lacked foresight in failing to include the slum dwellers in the construction process. The solution and installation are simple enough for the dwellers to understand and perform it themselves. As the architect stated in his interview (2014),

> …a product-based solution has spearheaded the NGOs' efforts rather than bringing the urban poor one step forward to help themselves to improve their living conditions.

4.3 Ramdev Pir Tekro Activity Centre Project in Ahmedabad, India

A community action project was initiated in 2004 by FootPrints Earth and Doshi & Associates for a slum community of 150,000 people (Ramdev Pir Tekro) in the city of Ahmedabad in India (Tiwari and Pandya 2014). Tekro dwellers have illegally squatted on the land, which belongs to the government, for at least the past 50 years. Many residents there, particularly young women and girls, work as rag-pickers in a waste recycling industry that provides the main economic driver for the community.

Using an ethnographic approach, consultants/researchers immersed themselves in the place and participated in the everyday lives of residents. Trust with local residents grew over a period of time prior to the community need assessment, which used observations and surveys, informal chats, and structured questionnaires. Conceptual design work for the Activity Centre, a multi-use, all-day community centre for various activities and age groups, followed. Detailed designs were discussed with the community. Building components were then tested, and new building prototypes using municipal waste (from the waste recycling industry) were designed (Figs. 4 and 5). Construction started in 2005 and the Centre was formally opened in 2006.

Aiming to create a horizontal power relation between different stakeholders, the project team conducted meetings, surveys, observations, informal chats, co-construction, community events, and awareness programs in a collaborative mode of inquiry. However, power resided primarily with the NGO owing to its long-standing relationship with the community. The role of local government was negligible, perhaps because greater local government participation would have raised bigger issues of landownership, infrastructure, and eviction.

To carry out the ethnographic investigation, the consultants shifted roles from designers to participant observers. By working simultaneously with the collaborative

Fig. 4 Recycled building components. *Source* Photos taken by Tiwari (2009)

planning model and ethnographic approach, the consultants engaged the community in co-designing and co-constructing, tapping into local issues, materials, tools, skills and labour. The tangible outcome produced 'by the people' and 'for the people' was the Activity Centre.

Capacity building for this project was defined as encouraging reciprocity, learning, and creative innovation (Innes and Booher 2003). Reciprocity was measured by the extent of participants' interdependence. The direct involvement of the local Tekro residents was critical in building a personal and collective identity. Residents were involved in making roof tiles and fly ash bricks, partitions and windows, collecting city solid waste, filling the glass and plastic bottles with fly ash, decorating and painting interior and exterior walls of the centre using their local designs and skills, and creating artwork in children's classes (Fig. 5) (Pandya 2008). As a result, the community members' sense of ownership and confidence in decision making grew stronger.

One of the most important outcomes of the project was the unquantified value of the economic empowerment achieved through the transfer of skills. Skilled and

Fig. 5 Residents' imprints. Photos taken by Tiwari (2009)

unskilled volunteers, together with construction professionals, encouraged mutual learning. Creativity and innovation emerged as spin-offs from the project. A model to reduce pollution and improve environmental factors through recycling municipal and domestic waste emerged. This model was implemented in three other centres. The importance of the local context was underscored in the project—local people, local resources, local tools and techniques—bringing in the bottom-up knowledge of the locale. The top-down facilitation and planning by the consultant/NGO team helped in a smooth run of the project, providing information, construction knowledge, mediation and facilitation.

5 Recurring Themes in Practice and Research: Participation, Bottom-up and Top-Down

There are many critiques of participation processes in community-led development projects. Most of these can be grouped into common themes:

- Participation is political
- Local power structures must be identified and incorporated into planning
- Low self-esteem of participants (owing to social, political and other factors) can result in participation avoidance
- The practitioner (consultants) must be accountable to the participants
- Researchers and practitioners must engage in ongoing self-assessment and reflection to improve practice

5.1 Participation Is Political

Participation can be influenced by political, social, cultural and spatial factors (Mubita et al. 2017). Servaes (2003) argues that participation should be independent of political agendas. However, Mohan and Stokke (2000) link politics with participatory practices to explain the nature of the movement. In linking politics to participation, two streams of development thinking are associated to theoretical reasoning. The first is Neo-Liberalism: emphasising market deregulation and institutional reforms to deliver development outcomes through civil society (essentially top-down approaches). The second is Post-Marxist: 'mobilisation of marginalised groups against the disempowering activities of both the state and the market' (Servaes 2003, 248) (bottom-up). Seraves (2003), Mohan and Stokke (2000), note that the 'local' should not be considered in isolation from the broader economic and political environment.

Tandon (2008) continues the discourse by suggesting that integrating models of participatory programmes into the democratic political system can be extremely successful in 'enhancing representative forms of democracy' (Tandon 2008, 293).

As seen in the YSUP case study, engaging the local NGO as local experts should, in theory, promote strong ties with the local community. Once the local leader was involved in the project, however, the political agenda was given priority, and this changed the dynamics and outcomes of the project. The community and the NGO were merely beneficiary and implementer (respectively). Political influences did help with the implementation, but community engagement was minimal and simply a means, not an end. The community should have been employed in the construction directly, with some technical and financial support, thus enhancing local skills to manage and maintain their own houses. Thus, the stated BSUP purpose, mission and objectives could have been achieved. In addition, subsidies could have reached more households, and the challenges of lack of finances and manpower could have been managed. Dissatisfaction resulting from the poor quality of construction and building materials could have been avoided.

5.2 *Awareness of Local Power Structures*

Power in this context simply means the ability to influence decision making (Mubita et al. 2017). As a critic of the effectiveness of participation, Dudley notes that effective participation may not occur owing to the threat to 'powerful vested interests'. It can either be used as a tool (often seeming 'paternalistic' and creating an 'us' and 'them') or a goal (where there is a true transfer of power).

Dudley's (1993) comments align with Parfitt's (2004) 'ends' and 'means' paradigm. Wakeford and Singh also note that the misguided agendas of invested interests and powerful players can influence participation. The misuse of participation to redirect power can include:

- Failing to target groups or minorities and only engaging those with power and vested interest (Parfitt 2004)
- Allowing powerful members of the community to 'hijack' the participation (Mubita et al. 2017; Parfitt 2004; Mathie and Cunningham 2003)
- Influencing participation via the power of the agency itself. '(The) power of aid agencies cannot be dispensed with in the simple way that he (Chambers 1994) envisages' (Parfitt 2004, 543)

Disenfranchised, poor, and marginalised individuals of a community may not participate in a meaningful way because they fear speaking out against the interests of powerful community members, or simply because of lack of trust in the facilitators. This can occur if local power structures are not identified and managed. In fact, they may be inadvertently reinforced via participatory exercises (Mubita et al. 2017). Voluntary participation is only voluntary insofar as power dictates.

In reflecting on the three Indian case studies, we can use the participatory approach and outcomes to position the projects against Arnstein's ladder of participation. The YSUP project sits on the fourth 'rung' of consultation. The process provided an opportunity for the 'have-nots' to hear and have a voice (Arnstein 1969), but with a level of 'tokenism' by not handing over control or influence in the project. The ICPP project sits between the fifth rung of placation, and the sixth rung of partnership (Arnstein 1969). The community was a major informer of the project and in some ways influenced the implementation of the roof module installations.

5.3 *Negotiate and Engage in Trade-Offs*

The third case study, Ramdev Pir Tekro, demonstrated greater empowerment of the community and this study therefore sits on the sixth rung of partnership. The community actively partnered in the development of the Activity Centre and the project outcomes.

5.4 Participant Self-esteem and Participation Avoidance

There are many reasons why individuals do not participate in community-led planning or participatory programmes. These reasons can be social, political or spatial. A major impediment to participation is the individual's perceived or real feeling of inadequacy (Nuttavuthisit et al. 2014; Long 2001). This supports Thornley's (1999) observation that community interests are fragmented. Participatory programmes are often based on the assumption that the poor are acutely aware of their needs and problems (Long 2001) and have a thorough awareness of themselves (Jacobson and Servaes 1999). Jacobson and Servaes (1999) insist, however, that facilitators should understand that individuals (including oneself) are not always centred, conscious, and ordered and can, in fact, be in constant movement between order and chaos.

IFAD, ANGOC, and IIRR (2001) introduce a new discourse about the hidden costs which community members pay (but not the facilitator). These include the time villagers contribute to the program by hosting the project team (e.g., tea and biscuits), facing the consequences of speaking out (caste and gender implications), the potential for domestic discord and loss of competitive advantage (when voicing skills to the open forum), and the risk of social posturing. The practitioner/facilitator must be acutely aware of all the factors which impact individuals and ensure they are fully available to participate and with minimal hidden costs.

5.5 Accountability and Self-assessment

To promote constant improvement and accountability, practitioners must create multiple opportunities for aid recipients to provide feedback and for practitioners to genuinely self-reflect on processes and outcomes of participatory projects/programmes. Botes and Van Rensberg (2000) identify the trend for reflection on projects to concentrate on these hard issues. Physical, financial and technological issues can be perceived as more important than the soft issues such as social impacts, community involvement, decision-making procedures, and the need to achieve a balance between product delivery and participation.

The YSUP project team surveyed beneficiaries who provided honest feedback. This feedback validated the perception that political biases and favouritism did influence the project. The feedback also focused on how the project could have been delivered more effectively and highlighted the values of decision making throughout the programme and skills development in the community.

6 Reflexivity: The Middle Ground Between Top and Bottom

While the top-down, bottom-up paradigm as a concept has been criticised, a consistent message from research and practice is that participation is essential. The authors are therefore encouraged to present a more dynamic, 'middle ground' approach to participatory planning.

An agency (i.e., an NGO, government agency, or other) which works to implement urban renewal/community development projects needs to create 'effective' participatory programmes. Practitioners, however, should avoid criticising participatory planning which may not demonstrate comprehensive bottom-up strategies (or reaching higher rungs on Arnstein's ladder of citizen participation) and should try to adapt to organisational constraints, while engaging in meaningful participation. It is possible to overcome the top-down/bottom- up paradigm by finding a comfortable middle space that meets the needs of the planning community, and this can be done through reflexivity.

Reflexivity, in social theory, is '(the) practice (that) involves intentional introspection for the purpose of improving a behaviour or process to achieve a desired outcome' (Wende et al. 2018).

The middle ground acknowledges the successes of a range of bottom-up approaches and uses these approaches as guides rather than manuals. Learnings are used as a starting point.

Adapting to the context and needs of locals requires new skills and a new approach. Rather than strictly controlling implementation programmes, professionals should be flexible and allow the programme to change to accommodate local needs (Craig and Porter 1997).

Many researchers encourage flexibility in the process of participatory projects (Parfitt 2004; Bhattacharjee 2001; Symes and Jassar 1998). Reflexivity is not just flexible process and delivery, but a consistent reflection on successes and failures, and improved processes. Many global case studies demonstrate that on-the-ground learning and reflections can change a project's direction. In their case study with PARC (Palestinian Agricultural Relief Committee), Symes and Jassar (1998) included participation, but it was in such a tight political context (under Israeli military occupation) that the needs of locals were not always responded to in a flexible manner. Bhattacharjee's (2001) social mapping project included asking a group of participants to draw a map of their village. The program was both interactive and interesting, but it was subject to a number of limitations. These limitations included a lack of time to thoroughly explore participation, locating participation in an upper caste house (therefore involving more upper caste villagers), language barriers, and the involvement of women.

In the Thai socio-political context, Nuttavuthisit et al. (2014) aimed to develop a few key mechanisms to 'enhance community development' and to manage the contextual inhibitors to meaningful, democratic participation. These studies demonstrate a reflexive approach to participatory planning that adapts to traditional bottom-up planning with top-down like implementation.

Parfitt (2004) recommends three key procedural adaptations to provide a more flexible and dynamic approach to participatory planning:

- Flexible, ad hoc, innovative learning units i.e., the 'ability to make ad hoc decisions when necessary' (Parfitt 2004, 549)
- Flexible budgeting
- Accountability to aid recipients

Based on the review of the three Indian case studies, the authors combine Parfitt's theory with our own key learnings on other significant aspects of participatory planning:

- Ongoing reflection and assessment (Wakeford and Singh 2008) and self- critical awareness (Parfitt 2004)
- Strong and effective communications within the project team (Tiwari and Winters 2016)
- Detailed contextual analysis continually informing decision making (Mubita et al. 2017; Nuttavuthisit et al. 2014)

The YSUP case study suffered from the inability of agencies and practitioners on the project team to monitor and assess the project against the intended capacity-building outcome. The project team found it difficult to comprehend the concept of capacity building. After-the-project impact evaluations are remarkably few, more result-oriented, and biased (Trivedi 2017).

The project team for the ICPP and Ramdev Pir Tekro case studies included time and flexibility in the planning for the learning phase (via their 'learning unit') for dialogue, exposure, and reflection while exploring the slum and its people. This bottom-up participatory design allowed for more organic, immersive learning. This resulted in a deeper understanding of the needs of households, home-based industries in the ICPP case, and local industry and waste management in the Ramdev Pir Tekro case. From the beginning, the Ramdev Pir Tekro project aimed to develop the skills and capacity in the community: skills to recycle nearly 20 types of waste materials into affordable building components for individual or mass-produced homes were developed. The programs in the Activity Centre included a women's centre, a co-operative store, and a crèche run by local grandmothers for slum children while their parents were at work. This became a self-sustained and convenient facility for the settlement.

The ICPP case study demonstrates the benefit of a flexible budget and implementation which allows local industries, skills and individuals to develop suitable solutions. The essential top-down support was also evident in Ramdev Pir Tekro project. This could have had wider impact on the everyday lives of the people if the local government had raised bigger issues of land ownership, infrastructure and eviction.

7 Conclusion

Participatory planning approaches, in general, are replete with design challenges. Participation in development projects has evolved as a response to failures in top-down approaches. Top-down development is traditionally characterised by the agency (be it local government, non-government organisation, or other) determining the needs of a community and the necessary actions to address those needs (Mubita et al. 2017). In the 1980s and 1990s, this method was criticised on a number of grounds, including neglecting to engage local communities (Mubita 2017; Nuttavuthisit et al. 2014), being ineffectual in delivering successful development programmes, and disconnecting individuals by making them feel incapable of 'taking charge of their lives and of the community' (Mathie and Cunningham 2003, 476).

Active participation of local communities in all stages of the project is essential for success: 'The people know their community and its issues; they have to live with the results, and can, want and have the right to participate' (Cities Alliance 2003, 21). The participatory strategy begins with a realistic needs assessment and becomes a precondition of each stage of the project (Tiwari 2009). Being involved in decision making empowers the residents to take responsibility for resolving their own problems. The NGO's networking role is important due to their long-standing local relationships.

A bottom-up approach does not encourage reciprocity, skill development and innovation without support from the top. Residents must be supported to obtain the skills they need to participate. They should be encouraged to express their views on local issues and make informed decisions about their priorities, and they should be treated as partners in the design and construction. Facilitating participation from a broad spectrum of residents through representation structures requires multi-disciplinary expertise, knowledge and skills. Top-down support is essential to reorient planning and financial measures to ensure resident involvement in activities that meet their needs and aspirations.

In this chapter, we have detailed a participatory planning approach that can achieve a balance between the bottom-up and top-down process approaches. A good example of this approach is the Ramdev Pir Tekro project. Craig and Porter (Craig and Porter 1997) outline two aims of development projects: Participation and effective Management. Participation incorporates unknowns through initiatives led by local people, while Management requires determined objectives (once again addressing the top-down bottom-up paradigm). Using a flexible approach that brings together participation and management through a hybrid model of top-down and bottom-up allows for design creativity, inclusive participation and a feasible implementation of the project to improve the building capacity of the community.

References

Albrechts L (2013) Reframing strategic spatial planning by using a coproduction perspective. Plann Theory 12(1):46–63

Arnstein S (1969) Ladder of citizen participation. J Am Inst Plann 35:216–224

Ashford G, Patkar S (2001) The positive path: using appreciative inquiry in rural Indian communities. International Institute for Sustainable Development, Winnipeg

Bhattacharjee P (2001) Social mapping at Thenganayakanahalli Village. Participatory Learning and Action Notes 2001 (41)

Botes L, Van Rensburg D (2000) Community participation in development: nine plagues and twelve commandments. Commun Dev J 35(1):41–58

Buehn A, Schneider F (2012) Shadow economies around the world: novel insights, accepted knowledge, and new estimates. Int Tax Publ Finance 19(1):139–171

Bushe GR (2011) Appreciative inquiry: theory and critique, pp 87–103

Chambers R (1994) Participatory rural appraisal (PRA): analysis of experience. World Dev 22(9):1253–1268

Cities Alliance (2003) Ann Rep

Cooperrider D, Srivastva S (1987) Appreciative inquiry in organizational life. Res Organ Change Dev 1:129–169

Cornwall A (2011) The use and abuse of participatory rural appraisal: reflections from practice. Agric Hum Values 28(2):263–272

Cornwall A, Pratt G (2011) The use and abuse of participatory rural appraisal: reflections from practice. Agric Human Values 28(2):263–272

Craig G (2007) Community capacity building: something old something new? Crit Soc Policy 27:335–359

Craig G, Mayo M (1995) Community empowerment: a reader in participation and development. Zed Books, London

Craig D, Porter D (1997) Framing participation: development projects, professionals and organisations. Dev Pract 7(3):229–236

Dudley E (1993) The critical villager: beyond community participation. Routledge, London

Elgin C, Oztunali O (2012) Shadow economies around the world: model based estimates

Gegeo D (1998) Indigenous knowledge and empowerment: rural development examined from within. Contemp Pacific 10(2):289–315

Healey P (1997) Planning—environment—cities. MacMillan Press Ltd., London

Innes J, Booher D (2003) Collaborative policy making: governance through dialogue. In: Hajer MW, Wagenaar H (eds) Deliberative policy analysis: governance in the network society, pp 33–59. Cambridge University Press, Cambridge

International Fund for Agricultural Development (IFAD), Asian NGO Coalition for Agrarian Reform and Rural Development (ANGOC) and International Institution of Rural Reconstruction (IIRR) (2001) Enhancing ownership and sustainability: A resource book on participation. IFAD, Philippines

Jacobson T, Servaes J (1999) Theoretical Approaches to Participatory Communication. Hampton Press Inc, Cresskill

Long CM (2001) Participation of the poor in development initiatives: taking their rightful place. Earthscan, London

Mashal (2011) Pune slum atlas. Pune, India

Mathie A, Cunningham G (2003) From clients to citizens: asser-based community development as a strategy for community-driven development. Dev Pract 13(5):474–486

Mohan G, Stokke K (2000) Participatory development and empowerment: the dangers of localism. Third World Q 21(2):247–268

Mubita A, Libati M, Mulonda M (2017) The importance and limitations of participation in development projects and programmes. Eur Sci J 13(5):238–251

Nanavaty R (2007) Exposure and dialogue programmes at Sewa. Participatory learning and action 57 Immersions, pp 26–29
Nuttavuthisit K, Jindahra P, Prasarnphanich P (2014) Participatory community development: evidence from Thailand. Commun Dev J 50(1):55–70
Pandya Y (2008) Manav Sadhana Crèche: revisiting the recycled. Vastu-Shilpa Foundation for Studies and Research in Environmental Design, Ahmedabad
Parfitt T (2004) The ambiguity of participation: a qualified defence of participatory development. Third World Q 25(3):537–556
Patel S (2013) Upgrade, rehouse or resettle? An assessment of the Indian government's basic services for the urban poor (BSUP) programme. Environ Urban 25(1):177–188
PMC (2013) Revised city development plan for Pune
Servaes J (2003) Approaches to development: studies on communication for development. UNESCO, Paris
Smith BC (1998) Participation without power: subterfuge or development? Commun Dev J 33(3):197–204
Symes J, Jassar S (1998) Growing from the grassroots: building participatory planning, monitoring and evaluation methods in PARC. Particip Learn Action Notes 1998(31):57–61
Tandon R (2008) Participation, citizenship and democracy: reflections on 25 years of PRIA. Commun Dev J 43(3):284–296
Thornley A (1999) Urban planning and competitive advantage: London, Sydney and Singapore. LSE London Discussion Paper no. 2
Tiwari R (2009) Urban crafting: making a connected city. REAL CORP 2009
Tiwari R, Pandya Y (2014) An ethnographic and collaborative model of inquiry: activity centre project in India. In: Tiwari R, Lommerse M, Smith D (eds) M2 Models and Methodologies for community engagement. Springer, Singapore
Tiwari R, Winters J (2016) The death of strategic plan: questioning the role of strategic plan in self-initiated projects relying on stakeholder collaboration. Int Plann Stud
Trivedi N (2017) Adopting collaborative planning for redevelopment of built environment as a means for capacity building of the urban poor. Doctor of Philosophy, Department of Urban and Regional Planning, Curtin University
Trivedi N, Tiwari R (2012) Collaborative dialogue and action for home-based work issues in Indian slum settlements. Reflect Built Environ Res 2(1):51–57
Wakeford T, Singh J (2008) Towards empowered participation: stories and reflections. Particip Learn Act Notes 58:6–9
Wende lML, Garney WR, Castle BF, Ingram CM (2018) Critical reflexivity of communities on their experience to improve population health. Am J Publ Health 108(7):896–901

Integrated Planning and Governance. Urban Regeneration Through Institutional Coordination

Institutional coordination refers to the capacity of making complementary the initiatives against inequality driven at different levels of government and in the different functional areas of a single level of government. The traditional urban planning approach and the "formalisation" of informal settlements has usually failed in the objective of incorporating informal settlements into the formal city. However, in the last decade, alternative planning methods have been incorporated taking into consideration the role of communities and neighbours. The improvement of institutional coordination is encouraged with the development of synergies between the initiatives promoted locally and nationally. The objective is to build the necessary critical mass for policies to have a transformative effect, which ensures the continuity of the policy beyond government cycles. The political agendas and the prioritization of resources happen generally in an unequal relation between municipalities and regions. The leadership at every level of the government is a key factor in creating a critical mass of local politics, which can be more dynamic and oriented with the demands.

Governance, Institutional Coordination, and Socio-Spatial Justice: Reflections from Latin America and the Caribbean

Clara Irazábal

Abstract Dysfunctional governance and institutional coordination are two of the aspects hindering improvements in socio-spatial justice in the Latin American and Caribbean (LAC) region. This is no minor issue in a region with one of the largest levels of informality and inequality in the world and where poverty rates are again expanding after showing encouraging retreats in the previous decades. Some of the governance challenges include unclear and unstable institutional structures for decision-making, insufficient mechanisms for citizen participation, and a lack of political will and appreciation for the benefits of planning. This chapter discusses governance and institutional coordination in LAC and suggests governance reform solutions for improving socio-spatial justice in formal and informal settlements by analysing current conditions and learning from cases of different cities and countries in the region.

Keywords Governance · Institutional coordination · Socio-spatial justice · Latin America and the Caribbean · Cities

1 Introduction

In any society, deficiencies and dysfunctionalities in governance and institutional coordination can sensibly restrict progress in socio-spatial justice. Conversely, good governance and institutional coordination of the kind suggested in this chapter[1] can have disproportionately positive effects in advancing socio-spatial justice. In Latin America and the Caribbean (LAC), while there is large variation in governance arrangements and performance among countries and within countries, and even among cities within a metropolitan or regional area, there is still ample room

[1]This chapter borrows from and updates Irazábal (2009c).

C. Irazábal (✉)
Department of Architecture, Urban Planning + Design (AUPD),
University of Missouri, Kansas City, USA
e-mail: irazabalzuritac@umkc.edu

O. Carracedo García-Villalba (ed.), *Resilient Urban Regeneration in Informal Settlements in the Tropics*, Advances in 21st Century Human Settlements,
https://doi.org/10.1007/978-981-13-7307-7_8

and acute need both to improve governance and institutional coordination and to realise their positive potential in improving socio-spatial justice.

This chapter discusses governance and institutional coordination in LAC and their role in the regeneration of formal and informal settlements, presenting an analysis of current conditions and offering examples of different countries and cities in the region organised according to: institutional frameworks for shaping plan formulation and implementation (national, regional, metropolitan, and local planning structures); impacts of governance on plan formulation and implementation; impacts of neoliberal regimes on planning and lower-income groups; and institutional arrangements between different levels of government and the role of decentralisation.

The chapter also offers normative recommendations for the improvement of governance in LAC, covering aspects such as: approaches through which planning has been integrated into government work; the governance of plan formulation and implementation; institutional arrangements and regulatory frameworks for effective plan formulation and implementation (metropolitan and regional planning, and international assessment and guidance of planning capacity and aims); and, finally, the roles that other actors—civil society and the private sector—play in plan formulation and implementation.

2 Institutional Frameworks for Shaping Plan Formulation and Implementation

The institutional framework for shaping plan formulation and implementation varies across the LAC region. It also varies depending on the level of government—national, regional, metropolitan, and local.

2.1 National Planning Structures

National planning in many LAC countries has benefited from greater institutionalisation in recent decades and is increasing its articulation with planning institutions at other levels of government as decentralisation advances in the region. In most LAC countries, there is a national agency in charge of planning, usually a national ministry of development, public works, planning, or a combination thereof. The implementation of national policies for development and territorial planning requires flexible and participative strategies allowing gradual and phased actions and coordination between a diverse set of actors, instruments, plans, and projects. This subsection presents a few examples of planning structures in the region.

In an innovative planning move, Brazil created the Ministry of Cities (*Ministério das Cidades*) in 2003, but this initiative has not been emulated elsewhere in the region. Other more conventional models contain ministries of planning and/or development, such as in Chile, Venezuela, Argentina, and Costa Rica. Lastly, some countries perform planning functions within other departments, such as those related to economic or social development, which is the case in Panama.

In Brazil, the Ministry of Cities' mission was 'to fight social inequalities, transforming the cities into more humanised spaces, through extending the population's access to housing, sanitation and transport'. Through cooperation with the Federal Economic Bank (*Caixa Econômica Federal*), the Ministry, created in 2003, worked with states, cities, social movements, non-governmental organisations, and the private sector. The creation of the Cities' Council in 2004 further added an instrument of democratic management to the National Urban Development Policy (NUDP). This council was a collegiate body with a deliberative and advisory nature for studying and proposing guidelines in the formulation and implementation of the NUDP, as well as following the execution of programs in housing, environmental sanitation, transport and urban mobility, and territorial planning. The Cities' Council was comprised of counsellors—representatives of segments of civil society and the three branches of government—with terms of two years, and state government representative observers, which also had to have Cities' Councils in their respective states. The institutional structure in Brazil may have been the most productive in LAC to advance urban planning, but it swang with the emphasis of its different leaders. Its mission was particularly threatened under the neoliberal focus of the current Bolsonaro administration, which changed the Ministry of Cities to the Ministry of Regional Development, potentially weakening its originally progressive urban focus.

In Panama, the former Ministry of Planning and Economic Policy (*Ministerio de Planificación y Política Económica*) was dissolved and some of its functions were adopted by the Ministry of Economy and Finances (*Ministerio de Economía y Finanzas*), created in December 1998 with the mission to formulate economic and social policy, administer and provide the resources for the execution of government plans and programs, and promote the greatest well-being of the population. It does not particularly focus on spatial territorial planning. Colombia's planning institutional trajectory is representative of a trend in the LAC area away from technocratic and centralised planning towards a more decentralised and participatory structure and a model of planning that aims to be more inclusive of, and responsive to, the communities in all stages of planning: plan formulation, implementation, and post-evaluation.

In Venezuela, the Planning and Development Ministry (*Ministerio de Planificación y Desarrollo*), created in 1999, was later called the Ministry of People's Power for Planning and Development (*Ministerio del Poder Popular para la Planificación y el Desarrollo*, MPPPD) to emphasise the socialist leaning of the Bolivarian Revolution. The Ministry's mission is to advise the President of the Republic, the National Assembly, and other government decision-makers in socio-economic development strategy formulation and instrumentation, plan promotion, creation of policies and projects compatible with the development strategy, and to facilitate coordination between the relevant groups and organisations for their execution. The Bolivarian political project in Venezuela has attempted to give greater power to the people to plan and implement projects in their communities. However, the inertia of traditional planning practices and models, the persistence and complexities of the old bureaucracy, the ambiguities and instabilities of new institutional and legal arrangements, government mismanagement and corruption, and the political opposition have been enormous challenges to surmount for this transition (Irazábal and Foley 2010).

In the Plurinational State of Bolivia, a country with a shorter tradition of planning, the Ministry of Developmental Planning, known previously as the Ministry of Sustainable and Environmental Development, was created in 2006 and has the mission of designing the guidelines for the governmental policies oriented at constructing a society and a state in which the Bolivians attain *bienestar*, i.e., "living well" and in harmony with "Mother Earth"—notions that come from the State's indigenous communities. The Ministry is responsible for planning and coordinating the country's integral development by means of the elaboration, coordination, and development of national economic, social, and cultural strategies, in relation to the other ministries, public departmental entities, locales, and social organisations representative of civil society. The Viceministry of Planning and Coordination has the function of planning the integral development of the country, contributing to the elaboration of the National Development Plan (*Plan Nacional de Desarrollo*, PND) with the collective participation of all sectors including farmers, microindustralists, small producers, and the private sector. It also formulates policies for the sustainable use of renewable and non-renewable natural resources, biodiversity and conservation of the environment, congruent with production processes and social and technological development.

The restructuring of Bolivian planning institutions and their intent is among the most innovative in the LAC region and deserves further commentary. The 2012 Framework Law of Mother Earth and Integral Development for Living Well guides the conception of national development policies, strategies, and programs in all sectorial and territorial scopes. Its aims are: to contribute to the country's transformation process; to disassemble the model of development conceived under colonialism and neoliberalism; to formulate and execute development while maintaining cultural sensitivity; to contribute to the construction of a new plurinational state promoting social communitarian development that equitably redistributes wealth, income, and opportunities; to develop a balanced coexistence of the state economy, the communitarian economy, the mixed economy, and the private economy; and to promote a new pattern of diversified and integrated development and the eradication of poverty, social inequality, and exclusion. Unfortunately, the government that tried to advance this agenda was toppled down in 2019 and the current one is hindering the implementation of this planning agenda. In addition, the recent intensification of mining practices as a major developmental strategy—what is been called neo-extractivism—is in contradiction with the stated objectives of the national plan, compromising both indigenous land rights and environmental conservation. Economic challenges in the last decade in the LAC region deriving from the Great Recession and the global drop in oil prices, among other factors, have promoted other countries in the region to also embark on neo-extractivist trends. The alarming economic crisis in LAC brough about by the 2020 pandemic is only making conditions worse.

Costa Rica's planning body is the Ministry of National Planning and Economic Policy (*Ministerio de Planificación Nacional y Política Económica*, or MIDEPLAN). MIDEPLAN formulates, coordinates, pursues, and evaluates the strategies and priorities of the government. MIDEPLAN is in charge of defining the route of national development and supporting the decision-making of the executive branch. Some of the main functions of the ministry are: to define both a medium- and long-term development strategy for the country; to draft the National Plan of Development, which articulates the government strategy in priorities, policies, programs, and actions; to coordinate, evaluate, and pursue those actions, programs, and policies; to maintain an updated evaluation of the national development's evolution so as to strengthen the processes of decision making and evaluating the impact of the government's programs and actions; to promote a permanent evaluation and renovation of state services (modernisation of public administration); and to stand by the application of government priorities in budget allocation, public investment, and international cooperation.

The National System of Planning was created in 1974 for the purposes of intensifying the growth of national production and promoting the best distribution of national funds and social services. The System is integrated with MIDEPLAN, the ministries' planning units or offices, and other public institutions. The National Plan of Development is created with participation from the ministers of state, managers of public institutions, representatives of international organisations, the diplomatic body, the Catholic episcopal conference, directors of universities, academics, mayors, deputies, government officials, and the media. Recent plans have included social policy, productive policy, environmental policy, institutional reform, and foreign policy. They respond to the following national goals: fight corruption in all the public sector's scopes of action; reduce poverty and inequality; increase economic and employment growth; improve the education system's quality and expand its coverage; stop the growth of crime and drug trafficking, and curb drug addiction rates; revert feelings of increasing insecurity within the citizenry; strengthen public institutions and order state priorities; repair and extend the country's transportation infrastructure; dignify foreign policy and improve Costa Rica's role in the world.

Planning institutions and practices in Costa Rica are advancing but not at the pace of development. In particular, pressures from globalisation—for example, for the instalment of international business parks and tourist-real estate developments and their related urban infrastructure—are affecting fragile ecosystems and overloading inadequate urban infrastructure (Irazábal 2018a). Planning legislation, institutions, and compliance are not keeping up with the pace of changes. This is also a common trend in the Caribbean nation states. However, the initiatives and institutions commented on here are, or at least were until recently, steps in the right direction towards overcoming political and institutional constraints on the promotion of planning in Costa Rica and other LAC countries.

2.2 Regional and Metropolitan Planning Structures

Many countries in the region are administratively divided into states, departments, or provinces with varying levels of planning independence, infrastructure, technical competence, and resources. Larger and wealthier states or departments are more likely to have planning institutions than smaller and poorer ones, and those institutions are better equipped and more technically proficient. For example, state planning institutions in northern Mexico are wealthier and better structured than in southern Mexico. State planning institutions in southern Brazil, on the contrary, are wealthier and more technically proficient than in northern Brazil. Similarly, planning institutions in mineral-rich states such as Zulia in Venezuela are richer than in poorer states, such as Apure. There are regions that lack planning institutions altogether, or whose planning is performed by central planning agencies, particularly those in capital cities of nation states.

Metropolitan areas are composed of municipalities or cantons with economic ties and joint planning and coordination of certain services. Metropolitan systems and problems cannot be fully addressed from a local perspective, since integrated visions across all cities in the region are necessary to congruently and effectively address the issues at both the metropolitan and local levels. There are some efforts to create metropolitan planning structures in some LAC regions, usually in the form of metropolitan councils of governments, but, where they exist, they lack sufficient power and do not supersede local governments; they play advisory, non-binding roles. Some important factors critically contributing to the quality decline of planning and governance in many LAC cities include institutional deficiencies and a lack of coordination of planning and governance at this metropolitan level.

The Main Mayoral Office or *Alcaldía Mayor* was a figure of Spanish colonial government in Latin America, from Peru to Nueva España (current day Mexico). The figure of the Alcaldía Mayor corresponds today to the main mayoral office in metropolitan Caracas, Bogotá, and other LAC cities. The Metropolitan Mayorship of Caracas, for example, coordinates and applies policies for development in the city's Metropolitan District. The Metropolitan District of Caracas is the City of Caracas' political unit based on a formula of administration and municipal organisation that is divided into two levels: a metropolitan level and a municipal level, with the latter being subjected to the metropolitan level. The Metropolitan Mayor of Caracas, also known as Greater Mayor, depends on the Government Council as the superior advisory organisation. This body is composed of the mayors of the municipalities forming the Metropolitan District of Caracas. The Metropolitan (Town Hall and *Cabildo*) is the legislative body of the District, composed of metropolitan council members. Five municipalities make up the Metropolitan District. Unfortunately, the political polarisation in recent years between the national government and the Metropolitan Mayorship of Caracas has made congruent metropolitan planning in Greater Caracas virtually impossible (Irazábal 2011).

The Metropolitan Government of Bogotá includes an Institute of Urban Development (*Instituto de Desarrollo Urbano*). It was created in 1972 and was designed to execute physical infrastructure projects, maintenance, and improvement actions to provide Bogotá's inhabitants with access to transportation and public space, improve their quality of life, and reach sustainable development. Participation mechanisms are implemented to generate public buy-in. It attempts to advance the concept of a city of the people and for the people, with a human environment that promotes the exercise of collective rights, fairness, and social inclusion. It envisions a modern, environmentally and socially sustainable city, balanced in its infrastructures, territorially integrated, economically competitive, and participative. There is a Developmental Plan Bogotá Better for Everybody (*Mejor para Todos*), 2016–2020. Bogotá's Plan of Territorial Ordering (*Plan de Ordenamiento Territorial*, POT) includes the bases for land-use policy for a period of 10 years in the matters of: controlled growth, renovation or conservation of urban structures, land assemblages for infrastructure development, open space, public facilities, affordable housing programs, and environmental considerations. The territorial model dictated by the POT establishes general systems for the urban structure: road systems, transport, aqueducts, basic sanitation, public facilities, and public space. The Institute of Urban Development executes the plans, programs, and projects related to road, transport, and public space systems.

2.3 Local Planning Structures

Most LAC countries have undergone recent decentralisation and devolution programs since the 1980s. This has led to a rise in local government-level planning. Yet, localities also have different institutional frameworks and capacities for plan formulation and implementation. Larger and wealthier localities generally have sophisticated planning structures, and in the case of capital cities (such as in Brasilia, Mexico City, Caracas, etc.), there are generally ill-defined boundaries between local and national planning institutions.

One of the most efficient local planning institutional structures in LAC is that of Curitiba, but there are also other municipalities in LAC that have performed relatively well in terms of planning. In 2004, the Municipality of Rosario, Argentina, through the Secretariat of Planning, initiated an agreement process with the purpose of developing a New Urban Plan that, with citizen consensus, would consolidate the City Project (Municipality of Rosario, nd). Components of the plan for Rosario included: the territorial organisation of the metropolitan region; proposals for transformation; city completion and extension policies; land-use planning, urbanisation programs, urban agreements, a public land bank; plan management; and promotion of housing and provision of services. The content of the plan revealed a strong spatial focus. In the area of housing, the plan addressed housing and its relation to structural projects, irregular settlement policies, housing policies and services coordination, financing funds, the territorialisation of housing policies and services, participation of the private sector, and infrastructure service provision priorities. In the area of recovery of public spaces and facilities, the plan considered public spaces as a supporting element of neighbourhoods and important to urbanisation as a whole, as well as public space system categories, rules of public space design, and the private intervention, management, and operation modalities of public spaces. The preservation of the environment and the built patrimony dealt with defining criteria for assessing the value of heritage; applied strategies in historic preservation; mechanisms and instruments to protect urban spaces and elements; urban environment preservation; consideration of waterways and river basins, including the multiple types of contamination and residue treatments; and the control of existing productive facilities. The reconstruction of mobility dealt with the urban network of transit systems and urban transportation nodes. Finally, the dimension of centrality dealt with the central city, new urban centralities, and centrality management in the urban plan.

The planning office of the municipality of Porto Alegre, Brazil is the Secretary of Municipal Planning (Municipality of Porto Alegre, nd), under which there is the Municipal System of Management of Planning (SMGP), which is in charge of the Master Plan of Urban Development and Environment of Porto Alegre (*Plano Diretor de Desenvolvimento Urbano e Ambiental de Porto Alegre*, PDDUA) and the plan's realisation and application in the city. The Managing Plan of Urban Development is the basic defining instrument in the city's development model and is composed of seven strategies, including urban structure, urban mobility, use of private land, environmental qualification, economic promotion, and production of the city. The basic principles of the proposed spatial model are: decentralisation of activities through multicentred policies; sociocultural provision of services; mixed-use development preventing displacement of people and vehicles; controlled densification; recognition of the informal city through social interest policies; and environmental qualification and structuring through patrimony valuation and production stimulation.

Urban centres in the Caribbean also face a wide variety of infrastructural and planning challenges. One of the key strategies that have been adopted by national governments in the region to address these challenges is the formation of urban development corporations, which can operate at local, regional, or national levels. These are governmental agencies with powers to facilitate urban regeneration in specific areas in Jamaica, Antigua and Barbuda, and Trinidad and Tobago. The Urban Development Corporation (UDC) was formed in Jamaica in 1968, the St. John's Development Company (SJDC) in Antigua and Barbuda was created in 1986, and the Urban Development Corporation of Trinidad and Tobago (UDeCOTT) was established in 1994. These bodies have far-reaching planning and development powers within specific areas. For example, the SJDC is able to acquire, manage, or dispose of lands and to lay out, construct, and maintain roads, buildings, public parks, piers, car parks, and other public amenities within specified designated areas. Jamaica's UDC is similarly empowered to carry out and/or secure the laying out and development of designated areas. These three urban development corporations are meant to act as developers in the public interest and as agents of modernisation. However, as manifestations of neoliberalisation, when these corporations have succeeded in effecting large-scale transformations in the urban landscape, this has often been achieved through a top-down development process exempted from planning regulations and showing little accountability to the residents of the cities.

There is much variety regarding local government, planning performance, and levels of democracy in LAC countries, including problems in the zoning classification and governance recognition and structure of low-income, self-built communities in LAC metropolises. These are often invisibilised or unevenly accounted for in formal plans and by formal planning institutions (Almandoz 2012, 2013, 2014). As a result, they often do not get integrated into governance and planning institutions and plans, furthering their physical deprivation and socio-political marginalisation. Policies like participatory budgeting, which started in Porto Alegre, Brazil and has spread to multiple cities in the Americas and beyond, have brought new levels of participation to self-built communities in some Latin American countries. Communal councils and communes have had a similar effect in self-built settlements in Venezuela. However, racial, class, and spatial stigmatisation of these settlements persists, which, together with accentuated drug trafficking and armed violence plaguing many of them, makes it hard for these communities to be fully integrated into formal systems of governance and participation.

2.4 The Case of Metropolitan and Urban Planning in Curitiba, Brazil

An example of efficient urban planning at the local level that has not effectively transcended to a metropolitan level is that of Curitiba, Brazil. To guide the discussions over the Plano Diretor of the Municipality of Curitiba, a commission was created in 1965 (*Assessoria de Pesquisa e Planejamento Urbano de Curitiba*, APPUC). As the work evolved, the planning team felt the need to transform the advisory commission into an independent public institution. The Institute of Urban Research and Planning of Curitiba (*Instituto de Pesquisa e Planejamento Urbano de Curitiba*, IPPUC) was created five months later (on December 1, 1965) as a political means for injecting flexibility and dynamism into the process. This municipal entity was able to avoid the bureaucracy of city departments, with the goal of 'changing the city's appearance, preparing it for the future' with functional technocratic planning under the guidance of the military regime.

The structure of the IPPUC was conceived such that its Administrative Council included representatives of all levels of government departments, thus establishing functional links with the other agencies. During the creation of the Institute, all of its members shared the same political inclination, had previously participated in the redrafting of the Plano, and were appointed by Jaime Lerner—an architect, urban planner, and savvy politician who had led planning in the city through its most important stages. The IPPUC was then further vested with authority over all of the other government agencies. Since 1966, having survived political changes and being transformed in the process, the IPPUC remains the major planning agency in Curitiba. Aiding in its mission, the succession of mayors and governors that followed were also committed to the realisation of the plan through a propitiatory institutional environment, nurturing and protecting the political will necessary to carry out the Plano (Irazábal 2005).

Despite attempts to give priority to metropolitan issues and metropolitan integration by recent mayors of Curitiba, the COMEC, Curitiba's Metropolitan Region Coordination (*Coordenação da Região Metropolitana de Curitiba*) and the Special Secretary of Metropolitan Affairs SEAM/ASSOMEC (*Secretaria Extraordinária de Assuntos Metropolitanos*, created in 1997), there is a major structural obstacle to those metropolitan plans. If patterns of urban development similar to those of Curitiba were implemented at the metropolitan level, the economic service provided by those municipalities to Curitiba could be diminished. Furthermore, the discussion of metropolitan issues has not yet transcended the elite in Curitiba, i.e., it has not been transformed into a socio-political problem and, hence, the powers-that-be have not been compelled to confront their constituency with proposals to address it. This is changing, however, as demonstrated by recent mayoral electoral campaigns, in which candidates felt some pressure to explicitly address the problems of metropolitanisation.

There are additional institutional constraints that make metropolitan planning difficult in Curitiba and other Brazilian (and LAC) cities. Only municipal and state governments in Brazil hold decision-making powers. The state legislature is the governing body for legally defined metropolitan regions. Therefore, metropolitan entities such as the aforementioned COMEC and SEAM/ASSOMEC are not decision-making bodies. There have been proposals presented for the creation of a metropolitan parliament with political power and a forum for municipal council members to meet and negotiate metropolitan issues. Vested political interests have resisted the proposals, however. Compounding the problem is the absence of political will to confront the power asymmetries among neighbouring municipalities for the creation or coordination of a metropolitan government. The possibility even exists for higher political costs to weaker municipalities, derived from confrontation among local governments over the potential benefits they can get from power redistribution. Instead, metropolitan fragmentation has resulted in new municipalities and the incorporation of neighbouring municipalities into the metropolitan region. This has made the attempts for metropolitan government and coordination even more difficult. In some cases, the fragmentations and incorporations are not functionally justified in terms of the urban dynamics taking place in the region; rather, they may respond to political-administrative interests.

Within this current system, the only other institute capable of taking on metropolitan planning leadership is the IPPUC. Yet, today, IPPUC's actions do not have the same broad impact they had during the early years of the design and implementation of its first Curitiba Master Plan (*Plano Diretor*), given that the major planning actions conceived in the Plano have already been implemented. In effect, there was a clear shift in the role played by IPPUC since Jaime Lerner's last term as mayor of Curitiba (1989–1992); from an emphasis on structural urban planning to isolated architectural and landscaping interventions of lesser impact.

Since IPPUC is a municipal agency, it also faces institutional constraints when it comes to performing metropolitan planning. These institutional problems are partially solved through some municipalities' (lacking technical capacity) contracting of IPPUC or professionals from this institution to perform consulting services in developing their municipal *Planos Diretores*. This occurs in a context where neither of the two metropolitan institutions have deliberating powers: the SEAM/ASSOMEC is a political council comprising the mayors of all the municipalities of Curitiba's Metropolitan Region (CMR), with the mayor of Curitiba as president; and the COMEC is a technical planning organ for the metropolitan region. Decisions for each municipality remain in the hands of individual mayors, who may or may not comply with metropolitan plans. Given the financial, technical, and political weight of the municipality of Curitiba within the metropolis, however, the metropolitan planning and managing processes are led and often times co-opted by the Mayor of Curitiba and planners of IPPUC.

The aforementioned considerations working against metropolitan coordination today result in metropolitan-level planning policies in Curitiba (and other LAC cities) lagging behind reality: they are not structuring the growth of the city-region, but merely trying to remedy some of the resulting dysfunctionalities. This is one of the major contradictions of metropolitan development in Curitiba. The situation is entirely opposed to that of the planning process of the municipality of Curitiba with the Plano Diretor in 1965, which envisioned a desirable city of the future and provided guidelines for achieving it. Transportation, for instance, was a major structural element in that vision. Today, however, there are still under-served neighbourhoods in the metropolitan area, even to the extent of lacking transit services altogether. While it was once relatively convenient for the central city to have the poor occupying the fringes of the metropolis, over time the benefits attained from such socio-spatial segregation have evolved into heavy burdens for the government in terms of providing transportation and other urban services to those fringe settlements. The unequal distribution of transit and housing in the metropolis has also encompassed other urban services and amenities (Irazábal 2006). Most parks are located in formal areas of the central city, often located at the borders of the municipality of Curitiba, providing the city some green buffering from the poorer metropolitan area. The major cultural centres are in the municipality of Curitiba as well.

The fact that the government and planning officials have given insufficient attention to (or have even hidden, on occasion) the problems of the metropolis, and in particular the unattended needs of the lower income sectors of the population, has had the counter-effect of making it even more difficult to deal with and provide solutions to those problems today. Some current social problems, including the deficit of affordable housing; deficient education and health services; and increases in crime, violence, homelessness, and unemployment, have sprouted in a neglected peripheral environment and have become critical signs of a deteriorating urban environment as a whole. Furthermore, following a neoliberal economic model, recent administrations are promoting the privatisation of education, health, housing, sanitation, and other (traditionally public) services (Irazábal 2006). Privatisation in these areas has progressed incrementally, with the approval of the City Council. In this process, numerous workers have lost their benefits and protections, as job market flexibilisation becomes the norm. The union movement has not been capable of maintaining the publicness of services, with profound repercussions for the populations residing in the peripheries. Unions and other interest groups are, however, very critical of the privatisation processes attempted by the government. As the welfare state model is dismantled, they protest that it is not being substituted by other mechanisms that can respond to the needs of the most disfranchised populations. Social movement groups organised around issues of housing, education, and health are also opposed to these trends. These groups, however, have thus far not been large, mature, resilient, and/or vociferous enough to effectively contest these processes (Irazábal and Angotti 2017) and are now facing more daunting challenges under a federal administration that is openly hostile to them and their causes.

3 The Impacts of Governance on Plan Formulation and Implementation

Variation can also be seen among the different institutional frameworks for plan formulation and implementation in LAC. It is common for the agencies in charge of plan formulation to be different and disconnected from those in charge of plan implementation. This negatively affects many plans, including those related to housing provision or settlement upgrading, which may not reach the implementation phase. A key issue is the trend toward decentralisation in LAC and its effect on planning. Decentralisation has generally empowered municipalities and often times mandated them to do planning. However, this mandate has not necessarily followed up with an adequate devolution of resources and with the technical capacity to allow municipalities to respond to these new responsibilities. Hence, the primary planning focus of many municipalities in the region is economic development in order to keep afloat and, if possible, to generate the economic and financial conditions to support a more comprehensive planning approach.

Governance challenges for plan formulation include unclear or unstable institutional structures for decision making, deficient technical expertise for data gathering and analysis, insufficient mechanisms for citizen participation, lack of political will or appreciation for the benefits of planning, short-term perspective among elected politicians, disruptions between consecutive terms in office of elected representatives (electoral cycles and politics), clientelism, patronage, corruption, and the prevalence of technocratic and/or incremental decision-making models. Complementarily, governance challenges for plan implementation include lack of funding, deficient fiscal management, corruption, bureaucratic inefficiencies (red tape, unclear procedures, uneven application of procedures, lack of clarity or predictability, complications and delays, deficient technical expertise), and political difficulties (deficient inter- and intragovernmental coordination, inter- and intragovernmental competition, lack of political support, political instability, and distrust, etc.) (Joseph et al. 2017).

One or more of these factors, to varying degrees, interfere with the experiences of plan formulation and implementation in LAC.

Plan formulation and implementation are heavily politicised in LAC. Party-based politicisation often trumps technical expertise and community input in plan making and implementation (Letelier and Irazábal 2017). Ideological bias, electoral politics and term limits, clientelism, patronage, populism, nepotism, and/or paternalism are common in cities and countries in the region. Plan formulation and implementation may also be affected by purposeful sabotaging among different institutions and levels of government, especially if they are political rivals. Weak systems for political accountability and legal impunity perpetuate these problems.

To affront those challenges, public sector reform programs implemented in LAC have placed an emphasis on efficiency, effectiveness, and economic development, including government financial management systems reform and public sector performance reforms. Despite persistent governance challenges, there are plenty of examples of countries and cities, such as some of the ones presented above, which are making progress in improving governance and making plan formulation and implementation more effective at all levels of government.

4 The Impacts of Neoliberal Regimes on Planning and Lower-Income Groups

The neoliberal notion that an unregulated market is the best way to increase economic growth, which will ultimately benefit everyone (the so-called 'trickle-down economics'), has not produced the results envisioned by governments that adopted it in LAC (Irazábal 2009a, c). Cutting public expenditures for social services like education and health care has effectively reduced the safety net for the poor. The justification offered for these social provision cuts is to save public funds, yet, paradoxically, government subsidies and tax benefits for businesses have been maintained or expanded. Deregulation promotes the reduction of any government rules that might diminish profits, including protecting the environment and workers' job safety. In parallel, the privatisation agenda proposes selling state-owned enterprises, goods, and services to private investors. This trend includes banks, key industries, railroads, toll highways, roads and bridges, electricity, schools, hospitals, and even fresh water. Usually done in the name of greater efficiency, privatisation has mainly further concentrated wealth in a few hands and made the public pay even more for its needs. Lastly, the replacement of the concepts of the public good and community with individual responsibility (furthered by 'enabling' approaches) has pressured the poorest people to find their own solutions to their lack of health care, education, and social security and has largely considered them to be 'lazy' if they fail.

In LAC, powerful financial institutions like the International Monetary Fund (IMF), the World Bank, and the Inter-American Development Bank have promoted neoliberalism. Neoliberal regimes have greatly impacted planning in LAC through the dismantling of the welfare state and the rolling-out of privatisation and reduction of government. The first clear example of neoliberalism at work took place in Chile, influenced by the University of Chicago economist Milton Friedman, after the coup against the popularly elected Allende regime in 1973. Neoliberal reforms in LAC countries have eliminated some of the traditional channels of participation and representation available to lower-income groups, restricting the voice of many in spite of democratic reforms. In Mexico, for example, neoliberal reforms have reduced political participation of the poor, resulting in the impoverishment of democracy. Other countries have also followed suit; however, some of the worst effects have been found in Mexico, where wages declined while the cost of living rose significantly after the implementation of NAFTA.

Despite laws to protect LAC's indigenous people's lands and cultures, the effects of globalisation and the application of neoliberal policies have been devastating on indigenous communities in LAC (Tovar-Restrepo and Irazábal 2014). Some of these groups do contest neoliberal trends. For example, the Mapuches, an often-overlooked indigenous group that constitutes between 4 and 10% of Chile's population, have directly challenged both private interests and the traditional concepts of state and nation, raising broad claims to collective economic and political rights.

Reactions to the negative impacts of neoliberalism on social well-being and equity sparked a wave of anti-neoliberal regimes in LAC, starting in the early 1990s. They, in turn, had an impact on planning reforms in Venezuela, Chile, Argentina, Brazil, Nicaragua, Bolivia, Ecuador, Uruguay, and Paraguay. Some of these countries (Chile, Argentina, Brazil) have had recent turns to the political right again, challenging gains in social provisioning and social planning that had contributed to the reduction of both poverty and inequality in the region. The drop in global oil prices and the rise of violence, political polarisation, corruption, poverty, and, in some instances, international economic sanctions and isolation (such as in the cases of Cuba and Venezuela) have further exacerbated crises and led to a loss of quality of life in the region.

5 Institutional Arrangements Between Different Levels of Government and the Role of Decentralisation

Since the level of urbanisation on the continent is large and continues to expand, and medium-sized and large cities are growing into metropolises and megalopolises respectively, the need for cross-sectorial and intergovernmental coordination and cooperation in the region is critical, especially for the regeneration of informally built areas. However, there is weak cross-sectorial and intergovernmental coordination and cooperation for planning in LAC. Again, conditions vary from city to city and country to country, depending on the size and wealth of governments, their level of technical expertise, their level of politisation, and their type of leadership and political will.

Cross-sectorial and intergovernmental coordination and cooperation should be an explicit mandate in the objectives of each government sector and level in charge of plan formulation and/or implementation. There should be a system of monitoring, evaluation, and accountability in place, and a system of incentives and disincentives for individuals and institutional performers. Yet, there are multiple disincentives for local governments to participate in cross-sectorial, intergovernmental, and regional planning, including: requiring more effort, time, and resources that are hard to come by; diminishing the self-determination and power of local governments; and potentially becoming less rewarding in terms of electoral politics.

Some strategies that can assist in redressing the disincentives for cross-sectorial, intergovernmental, and regional planning in LAC include: educating, and training when appropriate, elected and appointed representatives and public servants about the benefits, responsibilities, and know-how of cross-sectorial, intergovernmental, and regional planning, such that they can invest their efforts, resources, and political capital in it; educating civil society about the benefits and responsibilities of cross-sectorial, intergovernmental, and regional planning, such that citizens can request it from elected and appointed representatives and public servants; adjusting incentives, missions, and regulatory statutes of planning agencies (and/or creating new cross-sectorial, intergovernmental, and regional planning agencies as necessary), such that they are mandated and rewarded for engaging in cross-sectorial, intergovernmental, and regional planning; and adjusting electoral systems and terms, such that they can accommodate the requirements of cross-sectorial, intergovernmental, and regional planning.

There is also a need to critically review attempts in various parts of the world to achieve integrated, cross-sectorial government and improved intergovernmental relations with a focus on planning, while also assessing the extent to which these have been successful and can be adopted and adapted in different cities of LAC. This should include approaches to the different ways of linking land-use management (regulatory planning) and spatial planning with other policy arenas (housing, health, education, etc.), including the rehabilitation or upgradation of self-built settlements. Lessons can be drawn from both 'weak' and 'strong' experiences and from diverse methods for producing integrated municipal plans.

6 Approaches Through Which Planning Has Been Integrated into Government Work

There has been a long-standing prevalence of comprehensive spatial plans and land-use management (master planning) in LAC; however, there has been a growing trend toward other approaches: strategic planning, 'planning under pressure', endogenous planning (planning cooperatives, núcleos endógenos or self-sufficient towns), redistributive planning (agrarian and land reforms and proposals in Cuba, Venezuela, Brazil, Bolivia, Ecuador), participatory planning (communal councils in Venezuela,

Ecuador; participatory budgeting in Brazil and other LAC countries) and socio-spatial planning, such as health clinics and subsidised food markets in squatter settlements in Venezuela ("social missions") or the Favela-Barrio or slum-to-neighbourhood programs in Rio de Janeiro.

Strategic planning and 'planning under pressure' have been tried in Venezuela and have also been adopted in other LAC countries where political stability is not guaranteed, and time and resources are scarce. Strategic planning is a powerful tool for diagnosis, analysis, reflection, and collective decision-making around present tasks, and for developing guidelines to help institutions adapt with maximum efficiency and quality to the changes and demands imposed by circumstances. Strategic planning contributes viability in reaching objectives, such that the plan is always in constant evolution. It presents flexibility to propose situational changes. A high-level strategic framework should exist that allows for the establishment of a dynamic process of evaluation.

There are also a few LAC countries experimenting with socialist planning. The Cuban model has been around for six decades as of 2019, but the experiences in Venezuela, Bolivia, Ecuador, Chile, Argentina, Paraguay, and Nicaragua are more recent and moderate. These latter countries aimed to soften the neoliberal regimes that prevailed in the latter part of the twentieth century through the integration of social welfare programs. Even Cuba, to a certain extent, is trying to strike a balance between socialist and capitalist enterprises as strategies for economic survival in today's world.

Emphasis on physical or spatial planning has been an important factor in LAC planning. Even today, there are LAC cities whose recent planning activity has concentrated heavily on urban design interventions: examples include Curitiba, Rio de Janeiro, and Bogotá (Irazábal 2008; Del Rio and Iwata 2004; Del Rio and Gallo 2000; Sánchez and Broudehoux 2013; Berney 2011). Physical planning has been used to upgrade or beautify formal parts of the cities and, in some cases, it has been a central aspect in upgrading informal settlements as well. Some of these spatial plans have been criticised for their cursory attention to social considerations, while the best of them have indeed considered a more balanced approach between social and spatial planning, such as in the case of 'integral urban projects' and 'integral barrio rehabilitation' in Medellin, under the umbrella of 'social urbanism' (Angueloski et al. 2018).

In general, LAC countries and cities have been eclectic in their adaptation and syncretism of different planning models, including most particularly the rational or technocratic planning model, advocacy or equity planning, democratic planning, and participatory or collaborative planning. I even argue that what some of the social housing movements are doing in Brazil and other LAC countries through vacant land and building occupations—inducing the state, the private sector, and communities to allocate vacant and underutilised lands to social housing, thus redressing the national land/housing socio-historical debt—enacts restorative justice as a new and potentially productive planning mode in the region (Irazábal 2018b).

7 The Governance of Plan Formulation and Implementation

There should be built-in institutional structures and procedural mechanisms within planning agencies that streamline the linkages between plan formulation and implementation, facilitating interagency coordination and management. Recommended measures include a system of regular interagency meetings, mandated coordinating activities, making funding contingent upon coordination, integrating additional incentives, and the institution of a meta-participation and meta-collaboration monitoring through the figure of a planning interagency broker. Economic and financial solvency, relatively homogeneous societies, and relatively non-pluralistic political climates have facilitated the advancement of plan formulation and implementation in some LAC cities. Plan formulation and implementation were also centralised under some dictatorial and technocratic regimes in LAC and hence made easier (Irazábal 2004). Examples include the case of Curitiba's Plano Diretor, developed under a national dictatorship and an appointed technocratic local government; Caracas's transportation and urban infrastructure creation under Pérez Jiménez's dictatorship; Brazil's creation of Brasilia under Kubitschek; and the case of Chile's so-called 'economic miracle' and neoliberal planning under Pinochet. These examples constitute a challenge in public perception for the current democratic regimes, which are required to deliver noticeable results in more difficult and pluralistic economic and political climates.

In all cases, the prevailing political structure in countries can affect the formulation and implementation of plans. This is particularly the case for countries in transition (from militaristic to democratic regimes, such as in Argentina, Chile, Paraguay, Nicaragua, and Haiti) and from neoliberal to moderate capitalist or proto-socialist regimes (such as Venezuela, and until recently, Bolivia and Ecuador), where the fragmented, unstable, and transitioning political structures affect how plans are formulated and implemented. These challenges are particularly evident in the case of countries transitioning toward or from left-leaning regimes while maintaining old oligarchies and bureaucracies with different levels of resistance to change. Challenges are also observable in the case of countries recently coming out of divisive civil wars or dictatorial regimes, such as El Salvador, Guatemala, Nicaragua, Chile, Argentina, and Haiti. Distrust among people who belonged to opposing factions in the still-recent past hinders planning initiatives in the present and nurtures conflicting and often altogether opposite societal projects for the future (Joseph et al. 2017).

Apart from the constraints that may derive from the institutional framework underlying the planning process, other types of constraints may also hinder the effective formulation and implementation of plans. These include: shortage of skilled personnel; dearth of finance and the necessary support on the part of planning authorities to carry out basic functions; dysfunctional decentralisation whereby local governments have a governance mandate but lack the human or financial resources to comply with it; and lack of political will to implement plans. These factors are particularly present in small and medium-sized cities in the LAC region, which, paradoxically, are facing higher rates of growth than larger and more established cities.

8 Institutional Arrangements and Regulatory Frameworks for Effective Plan Formulation and Implementation

8.1 Metropolitan Planning

Ideally, LAC governments would strive to institutionalise metropolitan planning agencies. These agencies would plan for all localities in a metropolitan area with power over municipalities and systems to monitor, evaluate, technically support, economically incentivise, and exert pressure for implementation at the local level. These metropolitan planning agencies should be backed legally and financially by state and national government agencies. Perhaps the best example of a metropolitan planning agency in Latin America is that of Area in Medellín. In collaboration with municipalities and universities (Irazábal et al. 2015), this agency produces sophisticated metropolitan plans that contribute to improving sustainability in the region; however, it lacks binding power. Associations of governments with only planning advisory roles, such as the COMEC in Curitiba, Brazil, are inefficient models to attain effective metropolitan governance and planning. However, these institutions could be reformed and vested with deliberative and binding powers and new ones created where they do not exist and are needed. Growing small to medium-sized cities, where the need for metropolitan planning is not yet evident or urgent, should try to create metropolitan planning agencies at that stage, taking advantage of their more manageable scale and less complicated political climate than those existing in large cities.

8.2 Regional Planning

Most countries in LAC are facing regional planning challenges that transcend inter- and intra-municipal and state political boundaries; examples include shared coastal zones, river basins, Andean regions, or maritime regions. Adequate regional legislation and institutions are also needed for addressing the growing challenges to sustainability brought about by the recent greenfield and tourist development in specific areas of LAC (Angueloski et al. 2018), particularly in fragile forest and coastal ecosystems (Irazábal 2018a). Even when the legislation and institutions are in place to monitor preservation and development in LAC countries, deficient systems of accountability often allow for uneven application of monitoring, evaluation, and sanctions for non-compliance. In addition, unethical behaviour, corruption, deficient professional expertise, and lack of sufficient human, technical, and financial resources often compound the effectiveness of environmental and regional planning in the region. The emergent, complex, and large challenges brought about by climate change and migrations also demand urgent planning at multiple scales, from the local to the international (Irazábal 2010).

8.3 International Assessment and Guidance of Planning Capacity and Aims

In order to be more effective at envisioning and creating institutional arrangements for effective plan formulation and implementation in LAC, it would be helpful to act on the basis of an international assessment of the capacity for planning. The Royal Town Planning Institute (RTPI, an international organisation based in the UK) and the Commonwealth Association of Planners for the Global Planners' Network (GPN), with the support of the Lincoln Institute based in the US, developed an initiative to help planning organisations and planners around the world to assess their capacity to respond to the challenges or urbanisation, identify where they think the most strategic gaps in capacity are, and establish support networks that can help fill those gaps.[2] This tool (and others) can help us learn how different countries plan and manage human settlements and find out their priorities in planning these settlements. The tool is a web-based self-diagnostic application that can be reached in several ways.[3] The target respondents for the project are organisations or individuals involved in planning human settlements. The diagnostic tool develops modules for five different sectors: organisations representing planners; central government ministries and local government; NGOs and community groups; planner training organisations and academics; and the private sector. Whatever the processes and actors involved, international assessment and guidance of planning capacity and aims need to involve two-way, decolonial communication and negotiation channels.

Relevantly, the UN-Habitat released the New Urban Agenda,[4] adopted in 2016 at the United Nations Conference on Housing and Sustainable Urban Development (Habitat III) in Quito, Ecuador (United Nations 2017). It represents an internationally shared vision for a better and more sustainable future. This Agenda is tied to Goal 11 (Sustainable Cities and Communities)[5] of the Sustainable Development Goals put forth by the UN Development Programme and globally adopted to lead development until 2030. LAC cities would be well served by adopting these guidelines and instruments in their planning.

---

[2]www.lincolninst.edu.

[3]http://www.surveymonkey.com/s.aspx?sm=BVoAHze7i6IVKGrpW/zgdA==, visit the International page of the RTPI website www.rtpi.org.uk/rtpiinternational where there is a link, or use the short address http://tinyurl.com/2u2ovm.

[4]http://habitat3.org/the-new-urban-agenda/.

[5]http://www.undp.org/content/undp/en/home/sustainable-development-goals/goal-11-sustainable-cities-and-communities.html.

9 The Roles of Other Actors—Civil Society and the Private Sector—in Plan Formulation and Implementation

9.1 The Role of the Private Sector

The role of the private sector has been particularly important in capitalist societies, where much of urban life is determined by private decision-making, including individuals and corporations who make decisions that affect public planning. The role of the private sector is very powerful in the capitalist systems of LAC, and also in those that are striving to create a more regulated capitalist or socialist model (Portes and Roberts 2005).

Usually in LAC, either the economic and political elites overlap, or the economic elite exerts a large influence over the political elite. Even in non-participatory plan making, the economic elite is consulted and often benefited by the plans, e.g., with Curitiba's Plano Diretor of 1965. Often, local economic and political elites have their interests entangled with international elites and serve as their local agents. On a different scale, the exodus of refugees, economic and climate migrants, or retirees from some countries to other countries within LAC also affects land and housing markets in local communities, with the potential negative effects of displacements and the socio-spatial polarisation of local societies (Irazábal 2018a; Irazábal and Angotti 2017; Irazábal and Fumero 2012).

In order to make progress toward equitable and sustainable development in LAC, it is imperative for the state to maintain a central role in regulating and mediating the private sector while focusing on social provisionining and the public good. The state's mediating role should concentrate in three main areas to support and promote sustainable and equitable developmental practices: education for fostering a socially and environmentally mindful business environment; institutional and regulatory support (creation and implementation of a system of incentives, disincentives, controls, taxation, accountability, and regulatory predictability); and infrastructural support (allocating capital investments and financing and securing infrastructure provision).

9.2 The Role of Civil Society

Civil society is generally weak in Latin America, yet social movements have been able to stop or slow neo-liberal projects in several countries (Irazábal 2008). However, these movements are mostly reactive, and they often disintegrate after they get tokenistic results or a sense of urgent crisis dissipates. Nonetheless, civil society is expanding social learning and organizing not merely on a local level, but also transnationally in LAC (Irazábal 2014). Examples include the World Social Forum in Porto Alegre, Caracas, and other parts of the world as a reaction to the World Trade Organisation and other international efforts perceived as advancing an unsustainable and inequitable development agenda, disfavouring the developing world. Environmental groups in LAC are among the most active in the defence of fragile ecosystems (e.g., coastal areas) and communal agricultural land (e.g., ejidos in Mexico), and in the promotion of alternative developmental models, such as ecotourism (in Costa Rica, Peru, etc.) (Irazábal 2018a; Angotti and Irazábal 2017). Social justice is also an important agenda for social movements in LAC. The landless movement (*Movimento Sem Terra*) and roofless movement (*Sem Teto*) in Brazil, and political parties such as the Worker's Party (*Partido Dos Trabalhadores*) in Brazil (Irazábal 2018b) or Fifth Republic Movement (*Movimiento V República*) in Venezuela have been advocates of planning for social equity with varying degrees of consistency and effectiveness. The state should play a strong role in protecting and strengthening civil society. The state's supporting role should concentrate in three main areas: the provision of civic and technical education; support for the independent organisation, institutionalisation, and capacity-building of community groups; and the provision of funding and technical assistance (Letelier and Irazábal 2017).

10 Conclusion

Planning's purpose is the realisation of the public good and the maintenance and nurturing of the commons, i.e., the enhancement of sustainability and justice in places and communities. Even as planning operates through different scales—transnational, national, regional, metropolitan, local—and subfields—land use, transportation, housing, community development, economic development, urban design, environmental planning, transnational planning, historic preservation, disaster planning, climate change adaptation and mitigation, and others—all planning projects and processes should inch us closer to more just and sustainable cities (Irazábal 2009a, b, c, 2014). This, in turn, conveys the notion that socio-spatial justice can either be process- or product-driven, and ideally both. How planning in general, and planning for upgrading informal areas in particular, is performed and implemented (process) can help or hinder societies to become more democratic and participatory. In turn, the 'what' of planning (product), i.e., the substance and effect of planning, should always tend towards justice and sustainability, redressing socio-spatial and environmental inequalities.

In LAC, the challenges that persist in governance and institutional coordination restrict progress in socio-spatial justice. This is no minor issue in a region with one of the largest levels of informality and inequality in the world and where poverty rates are again expanding after showing encouraging retreats in the previous decades. Conversely, improvements in governance and institutional coordination, of the kind suggested in this chapter and more, can have disproportionately positive effects in advancing socio-spatial justice in the region. This still-unfulfilled promise should prompt LAC governments to strive for governance reform, and communities to demand it.

References

Almandoz A (ed) (2012) Caracas, de la metrópoli súbita a la meca roja. Olacchi, Quito (Almandoz A (ed) (2002) Planning Latin America's capital cities, 1850–1950. Routledge, Londres y Nueva York

Almandoz A (2013) Modernización urbana en América Latina. De las grandes aldeas a las metropolis masificadas. Instituto de Estudios Urbanos y Territoriales, Universidad Católica de Chile, Santiago

Almandoz A (2014) Modernization, urbanization and development in Latin America, 1900s–2000s. Routledge, Londres y Nueva York

Angotti T, Irazábal C (2017) Introduction: planning Latin American cities: dependencies and 'best practices.' Lat Am Perspect Special Issue 2: Plann Latin Am Cities 44(2):4–17

Angueloski I, Irazábal C, Connolly J (2018) Grabbed landscapes of pleasure and privilege: socio-spatial inequities and dispossession in infrastructure planning in Medellín. Int J Urban Reg Res (Forthcoming)

Berney R (2011) Pedagogical urbanism: creating citizen space in Bogota, Colombia. Planning theory

Del Rio V, Gallo H (2000) The legacy of modern urbanism in Brazil. Paradigm turned reality or unfinished project? Docomomo J

Del Rio V, Iwata N (2014) The image of the waterfront in Rio de Janeiro: Urbanism and social representation of reality. J Plann Educ Res

Irazábal C (2004) A planned city coming of age: rethinking Ciudad Guayana Today. J Latin Am Geogr 3(1):22–51

Irazábal C (2005) City making and urban governance in the Americas: Curitiba and Portland. Ashgate, Aldershot

Irazábal C (2006) Localizing urban design traditions: gated and edge cities in Curitiba. J Urban Des 11(1):73–96

Irazábal C (ed) (2008) Ordinary places, extraordinary events: citizenship, democracy, and public space in Latin America. Series: Planning, history and environment, 2nd edn. Routledge/Taylor & Francis Group, New York/London

Irazábal C (2009a) One size does not fit all: land markets and property rights for the construction of the just city. Int J Urban Reg Res 33(2):558–563

Irazábal C (2009b) Realizing planning's emancipatory promise: learning from regime theory to strengthen communicative action. Plann Theory 8(2):115–139

Irazábal C (2009c) Revisiting urban planning in Latin America and the Caribbean. One of eight regional studies prepared as inputs for the seventh issue of the Global Report on Human Settlements. planning sustainable cities: Policy Directions GRHS. 2009, United Nations Human Settlement Programme. https://unhabitat.org/wp-content/uploads/2010/07/GRHS2009RegionalLatin AmericaandtheCaribbean.pdf

Irazábal C (2010) "Retos Urbano Ambientales: Disturbio Climático en América Latina y el Caribe"). One of five Conceptual Documents prepared as inputs for the first issue of the State of Affairs in Cities of Latin America and the Caribbean; in Spanish, Estado de las Ciudades de America Latina y el Caribe. United Nations-HABITAT Regional Office for Latin America and the Caribbean

Irazábal C (2011) Muddling through socialist planning in Venezuela: reflections from Caracas' experiments. In: Proceedings of the association of collegiate schools of planning (ACSP) 2011 conference. Salt Lake City, Utah. 16 Oct 2011, p 277

Irazábal C (ed) (2014) Transbordering Latin Americas: liminal places, cultures, and powers (t)here. Series: Routledge research in transnationalism. Routledge/Taylor & Francis Group, New York/London

Irazábal C (2018a) Coastal urban planning in 'The Green Republic': tourism development and the nature-infrastructure paradox in Costa Rica. Int J Urban Reg Res (Forthcoming)

Irazábal C (2018b) Counter land grabbing by the Precariat: housing movements and restorative justice in Brazil. Urban Sci 2(2):1–18

Irazábal C, Angotti T (2017) Introduction: planning Latin American cities: housing and citizenship. Latin Am Perspect Special Issue 3: Plann Latin Am Cities 44(3):4–8
Irazábal C, Foley J (2010) Reflections on the Venezuelan transition from a capitalist representative democracy to a socialist participatory democracy: what are planners to do? Latin Am Perspect 37(1):97–122
Irazábal C, Fumero G (2012) El Movimiento Ocupa Wall Street: Lecciones de Movimientos Latinoamericanos y de Derechos de los Inmigrantes en EEUU, Urban NS03, 141–153
Irazábal C, Mendoza-Arroyo C, Ortiz Arciniegas C, Ortiz Sánchez R, Maya J (2015) Enabling community-higher education partnerships: common challenges, multiple perspectives. Curr Opin Environ Sustain (COSUST) 17(C):22–29
Joseph S, Irazábal C, Désir AM (2017) Trust and hometown associations in Haitian post-earthquake reconstruction. Int Migr 1–29
Letelier F, Irazábal C (2017) Contesting TINA: community planning alternatives for disaster reconstruction in Chile. J Plann Educ Res (JPER), 1–19
Ministério das Cidades, Brazi.l. http://www.cidades.gov.br/
Ministerio de Economía y Finanzas, Panama. http://www.mef.gob.pa/es/Paginas/home.aspx. Ministry of National Planning and Economic Policy MIDEPLAN, Costa Rica. https://www.mideplan.go.cr/
Municipality of Porto Alegre. http://www2.portoalegre.rs.gov.br/portal_pmpa_novo/
Municipality of Rosario. https://www.rosario.gob.ar/web/
Portes A, Roberts BR (2005) The free-market city: Latin American urbanization in the years of the neoliberal experiment. Stud Comp Int Dev 40(1):43–82
Sánchez F, Broudehoux A (2013) Mega-events and urban regeneration in Rio de Janeiro: planning in a state of emergency. Int J Urban Sustain Dev 5(2):132–153. https://doi.org/10.1080/19463138.2013.839450
Tovar-Restrepo M, Irazábal C (2014) Indigenous women and violence in Colombia: agency, autonomy, and territoriality. Latin Am Perspect 41(1):41–60
United Nations (2017) New urban agenda

Community-Driven Forms of Governance in Thailand. City-Wide On-Site Upgrading of Informal Settlements

Oscar Carracedo García-Villalba

Abstract This chapter reviews the key aspects related to governance in Thailand, addressing the upgrading of informal settlements at the national level from a community-driven perspective. It presents the findings of research undertaken on two pioneer case studies under the Baan Mankong Programme. More specifically, the chapter examines the on-site upgrading governance, policies, and planning approaches that contribute to the development of economic, social, and physical improvements in informal settlements. It also evaluates to what extent the informal dwellers have seen improvements in their living conditions as a result of the implementation of the Baan Mankong Programme. The article elaborates on the contributions to the positive development of the low-income communities that can be extracted from the experience in Thailand and suggests valuable lessons for other cities going through similar processes, pointing out some challenges for the future development of on-site upgrading policies.

Keywords Informal settlements · Slums · Community-driven upgrading · On-site upgrading · Baan Mankong programme · Affordable housing · Thailand

Thailand, like most of the countries in Asia, is facing a fast urbanisation process. In recent decades, the country has made significant progress in economic development and poverty reduction. In the mid-1980s the Thai economy consolidated its shift from an agricultural economy to a manufacturing-based open economy, and in less than 25 years, the country shifted from a low-income country to an upper-middle-income one (Bhatkal and Lucci 2015).

As in most fast-developing economies, population growth and shifts in the economic structure from a primary and industrial economy to a tertiary one are driving a migration process from rural areas to cities. Cities and urban areas gather

O. Carracedo García-Villalba (✉)
Director Master of Urban Design, Director Designing Resilience in Asia International Research Programme, Department of Architecture, School of Design and Environment, National University of Singapore, Singapore, Singapore
e-mail: oscar_carracedo@nus.edu.sg; omc@coac.net

O. Carracedo García-Villalba (ed.), *Resilient Urban Regeneration in Informal Settlements in the Tropics*, Advances in 21st Century Human Settlements,
https://doi.org/10.1007/978-981-13-7307-7_9

most of the financial activities, commerce and services. Therefore, they become attractors for the migrant population because they offer new opportunities for the low-income community in the search of better job opportunities and services such as health and education.

1 Process of Informality

The Bangkok Metropolitan Region (BMR), as the main urban area and economic centre in Thailand, is the preferred employment destination for these rural-urban migrants. An increase in urbanisation has accompanied economic growth. Between 1990 and 2010, the population living in urban areas in Thailand increased by 15 percentage points, from 29% in 1990 to 44% in 2010 (World Bank 2014), and estimations say that in the 2020 census we will see that Thailand has become urban, with more population living in cities than in rural areas. In this scenario, Bangkok is the primary destination for labour migrants since the 1960s, which has contributed to its growing population (UN-Habitat 2003).

The process of informality in Thailand, and specifically in the Bangkok Metropolitan Region, is not new or unknown. Like in many other countries, the rural-urban migration process, coupled with the lack of effective policies, plans and affordable housing provisions for low-income residents, contributed to the inability to absorb the influx of the incoming rural population. If we add the limited budget available for these purposes and the lack of capacity and motivation of the Thai government (Senanuch 2004), this lays the perfect foundation for the urban spread of informal and unplanned settlements.

Still today, limitations in access to reliable data make the measurement of the magnitude of slums[1] a very tough, challenging and imprecise task. Several authors have tried to estimate the informal and poor population in the past. The first comprehensive survey of informal settlements in Bangkok and surrounding provinces was conducted in 1985, finding a total of 1011 informal settlements (845 slums and 166 squatter settlements), with an estimated population of 1 million people (Pornchokchai 1985). In 2000, reports from the NHA show that, in Bangkok alone, there were 1208 informal settlements with 243,204 households, compared to the 943 informal settlements and 173,890 households in 1985 (Viratkapan and Perera 2006). Boonyabancha (2004) also reflects the limitations of data. She explains that the National Housing Authority estimates that 13% of Thailand's population, approximately 105,000 families at that time, were living in 3500 informal settlements. However, she also highlights that there are reasons to believe that these figures are lower than reality, since surveys conducted found more people living in low-income settlements than the official statistics. Also, in 2004, the Thai government launched a nationwide poverty eradication scheme that started with a registration program. The

[1]The article will refer to slums, crowded communities, urban poor communities, and squatter settlements in Thailand as "informal settlements".

initiative was targeted to find the major problems that, according to the poor population, the scheme should tackle. This voluntary registering process lasted for three months, and a total of 7,897,852 persons registered (World Bank 2014; Senanuch 2004). Although, in this case, as Prime Minister Thaksin stated, the objective of the registration was to find how many people needed help and of what kind, and therefore the quantification might not be perfect since some people could have decided not to register, the figure can be taken as a reference indicating the relevance of the informal settlements and settlers.

Although there are no current statistics or data regarding the population living in informal settlements in Thailand or Bangkok, what can be affirmed is that there are some unique characteristics in the Thai case compared to other cities. On the one hand, as argued by Pornchokchai (2003), Bangkok has a significant number of informal dwellers (65.3%) who were born in the actual informal settlements in the city. With regards to the rest, rather than being considered as urban-to-rural migrants in their majority, they should be defined more precisely as rural-to-rural migrants, because of their temporary status related to seasonal migration following job opportunities. In this sense, it has been observed that most of the long-term permanent rural-to-urban migrants settle outside Bangkok's municipality, in the Metropolitan Region, where more affordable prices and possibilities to settle can be found.

On the other hand, the second aspect that makes the case of informal settlements in Thailand unique comes from the more intangible aspects of spiritual culture. As Yap and De Wandeler (2010) explain, people do not organise to invade vacant land; instead, they prefer to seek oral or written consent from the owner to occupy the land and build their houses. The emphasis on harmony that the Thai culture entails tries to avoid conflicts. Landowners with empty plots usually accept the temporary occupation of their land by those who cannot afford to buy it, as long as the arrangement does not affect private interests and can be terminated immediately once it is required and in the same terms and spirit of harmony that it was provided, without protest or any kind of compensation. In exchange for the provision of free- of-charge land, settlers pay a small rental fee. The fact that any empty plot or small parcel of land in Bangkok can be occupied by these informal settlements, even in between high-income residential areas, creates a very scattered model that makes their control and quantification a rather difficult task.

2 Forms of Governance

Historically, public institutions in Thailand have had a prevailing role in determining the economic and social development of the country. For the last 60 years, the government has used the National Economic and Social Development Plans (NESDP) to set the strategic approaches, the development priorities, and the economic resources (Teokul 1999). NESDPs are the documents in which, every five years, the government formalises and expresses the political vision by depicting the national economic and development strategy. The documents, currently in their 12th edition, lay out the development agenda, actions and flagship projects to prepare the society and the economy in Thailand for future challenges (NESDC 2016). The detailed analysis of these documents and the steps that national and local governments followed after them offer a clear picture of the different stages in the form of governance that Thailand has gone through.

2.1 From Slum Clearance to Slum Upgrading. Integrating Physical and People-Centric Development

NESDPs are documents that determine the government's urban poverty alleviation policies for both urban and rural areas. In this sense, the first steps in the history of slum upgrading in Thailand were taken in 1964 under the first NESDP (1961–1966) when the Central Office for Slum Improvement (COSI) was created under the Bangkok Municipality. Although this agency took actions related to slum clearance, and it was responsible for prevention and demolishing of slums in Bangkok, it also initiated the pilot project for the community of Bonkai as a slum improvement program, the first of its type, providing physical and infrastructure improvements (Senanuch 2004).

A few years later, in 1973, under the third NESDP (1972–1976), a new stakeholder, the National Housing Authority (NHA), was created as a government agency responsible for slum upgrading and providing low-income housing, taking over COSI's role. The approach that NHA initiated to confront the social issues of informal settlements took a particular focus on housing policies. Intending to develop large stocks of housing that could solve the problem of informality, NHA established the Office of Slum Upgrading (OSU) in 1977, during the fourth NESDP (1977–1981). Reaching beyond COSI's area of action, it was meant to have a broader influence in Bangkok and other regions. Although the approach to slum upgrading taken by the office was made through small-scale projects, NHAs large-scale vision precluded the inclusion of communities' needs (Bhatkal and Lucci 2015). To a certain extent, we could say that this was a step towards nationalisation and centralisation of the efforts to tackle the problem of informality from a broader perspective.

The confirmation of the shift from slum clearance to slum upgrading took form officially with the 1979–1982 NHA's Master Plan for Slum Upgrading. The plan established measures for the physical and environmental improvement of slums, in addition to supporting income generation for neighbours. Despite the efforts made through the plan, in 1986 the government decided to create the Bangkok Metropolitan Administration (BMA) as part of the fifth NESDP (1982–1986) after realising that the number of slum dwellers was still increasing dramatically in Bangkok (Senanuch 2004). The new agency took over from NHA the responsibility of implementing slum upgrading projects in Bangkok, returning to a decentralised approach 13 years after the dissolution of COSI. The new policies integrated the physical interventions with a people-centric approach, as the government wanted to reflect the changes that

Thai society was undergoing, opening and transforming rapidly towards an active, responsible role in solving the concerns of urban low- income groups through the establishment of many community-based organizations and NGOs.

2.2 A New Type of Governance. From Government-Driven to Community-Driven Development

In 1992, BMA in Bangkok created the Community Development Department. The Department was responsible for the construction and improvement of housing and informal settlements, and it was also in charge of encouraging income-generation activities and promoting people's organisations in informal communities (BMA 2001). The same year, as part of the seventh NESDP (1992–1996), the Urban Community Development Office (UCDO) was created under NHA. The establishment of these two new public agencies constituted a defining moment in the history of slum upgrading in Thailand and specifically in Bangkok, as they represented the first steps in a new type of governance that actively integrated the low-income population.

At that time, the benefits of the slum upgrading approach, because of the low-cost investments, were publicly recognised (Senanuch 2004). However, the country's economic growth in the 80s and early 90s was not reflected in an improvement in the living conditions of low-income residents. In many cases, they deteriorated even further because of a lack of public funds and an excessive reliance on private owners to solve the problem of informality that resulted in an abandonment of the communities. This situation evolved into the understanding that a more participatory model was needed to support the low-income population. With this purpose, UCDO, whose Board was participated by representatives from government and community organizations, was provided with an initial capital of approximately US$40 million to develop an Urban Poor Development Fund. The Fund should extend loans to community-based savings groups formed by low-income community groups, NGOs, or agencies to generate income, purchase land, and build or improve housing (Boonyabancha 2003). In that sense, UCDO was not intended to be an implementing agency, like NHA.

As stated by the Community Development Organizations Institute, through the fund low-income people could organise, start savings groups, build their own management systems, initiate housing development projects and settlement improvements, and generate income to access financial support for these projects from the UCDO fund, in the form of collective loans and some grants. This kind of institutional arrangement, in which poor communities take the lead in decision-making and implementation, and the public funds support the communities with flexible finance formulas, were new to everyone (CODI 2019).[2]

[2]Community Organization Development Institute (CODI). https://en.codi.or.th/community-network-activitys/codi-network-activities/.

This innovative approach through community savings groups proved to be very effective. Access to loans for low-income residents not only reduced vulnerability, but it also helped communities to learn how to develop and manage their resources and to find the most efficient ways to maximize the impact of their actions with the lowest costs while addressing the targeted problems. However, as the interest from communities in this upgrading approach increased and the number of community-based saving groups grew, UCDO started facing a problem of scale, which led to a renovated program in the development fund. The difficulties in supporting individual groups initiated a new system which consisted in linking different community-savings groups in the form of community-networks so that loans could be provided to both individual communities or community-networks, which, in turn, would sub-lend to their members. This innovation had a profound impact on the community-led process and in the mechanisms to make funds available for low-income groups (Boonyabancha 2003).

By 2000, eight years after its inception, the results and impact of UCDO's actions through community-based saving groups were impressive: 950 groups had been formed in 53 provinces; loans and technical support provided to 47 housing projects involving 6400 households; 68,208 families in 796 communities benefitted from grants for improvements in infrastructure and living conditions; and more than 100 community networks were established (Boonyabancha 2002).

It was also in 2000 when another crucial step towards strengthening the role of community-based organisations was taken. That year, under the eighth NESDP (1997–2001), the government merged UCDO with the Rural Development Fund to form the Community Organization Development Institute (CODI), one of the main actors in the transformation of informal settlements to this day. CODI also took over the programs and responsibilities from UCDO, but it mainly constituted a significant change in the governance structure. As a special project under the National Housing Authority (NHA), UCDO had no legal status of its own, and all its regulations had to be under NHA. But as a distinctive new public organization category of government institution, created as part of governance reform efforts, CODI had more opportunities and more independence to operate. Thus, while UCDO was an initiative by NHA, CODI was set up as an independent legal public agency provided with more freedom and flexibility. As explained by CODI, this means that "as a public organization, the government funds which CODI receives for its work come under a particular fiscal category called 'General Subsidy', which allows for greater flexibility in how the funds are used. This is extremely important for Thailand's people's processes because it means CODI can direct those resources directly to communities on the ground."[3]

We can conclude by saying that, with the establishment of CODI, the government demonstrated a change in the emphasis on economic growth, shifting toward a decentralised and people-centred development. CODI was able to provide links and

[3]Ibid., note 2.

network collaborations between urban and rural areas to form mixed community-based groups. In this sense, the UCDO program approach was given continuity through community-driven management, saving groups, and loans, but now increasing the number of participants with over 30,000 rural community and urban organisations (Boonyabancha 2003) with stronger community participation and a shared decision-making process. As CODI said, it represented a first stab at decentralising a portion of government development resources to a people-driven process.[4]

3 Affordable Housing Programmes and Urban Planning Policies in Thailand

In 2003, it was estimated that the number of households and urban low-income communities living in insecure housing conditions was 1.14 million (CODI 2005). Despite the differences between this figure and those mentioned before, the number was significant enough for the government to justify the initiation of two new national programmes to address the housing problems among low-income groups and provide secure housing from two different perspectives. The two new housing development programs were the Baan Eua-Arthorn (BEA) ('We Care Housing') and the Baan Mankong ('Secure Housing'), the latter of which is still ongoing.

3.1 'We Care'. The Baan Eua-Arthorn Programme (BEA)

The Baan Eur Arthorn Programme (BEA) is an initiative by NHA established under the ninth NESDP (2002–2006). BEA does not address informal settlements directly. Instead, it suggests tackling the problem from a tangential approach by building new affordable housing units for low-income residents. In this sense, essentially, BEA is a large-scale mass-housing estate programme whose main objectives are to provide secure land tenure and affordable and appropriate standard housing, equipped with the necessary infrastructure to respond to the lack of housing for the low-income population in both urban and rural areas (National Housing Authority (NHA) 2005) (Fig. 1).

The programme was very ambitious and targeted to build approximately 600,000 units—477,000 in the BMR, and 123,000 in other provinces—by the end of 2007 (NHA 2002). It included a two-fold approach: while NHA was in charge of construction and implementation of the housing, the Government Housing Bank (GHB) was responsible for providing affordable housing mortgages. 17% of the cost of each unit (US$2650 out of US$15,600) was subsidized by the government, resulting in a selling price of US$13,000 (390,000 Baht). GHB fixed an interest rate of 4% for

[4]Ibid., note 2.

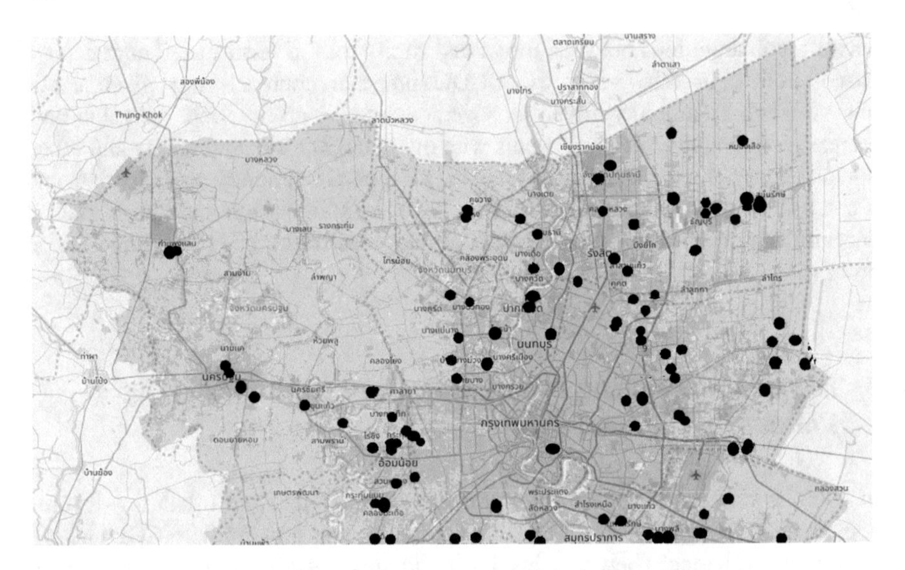

Fig. 1 Location of the Baan Eur-Arthorn projects in the Bangkok Metropolitan Region. *Source* Author from NHA 2014 data

the first three years, 5% for the fourth and fifth year, and a floating rate after that. After studying the income of the low-income population, NHA came to the conclusion that families would be able to use 20% of their household income to pay the loans on monthly instalments so that the housing would be affordable for those with monthly earnings of US$330 to US$500 (10,000 to 15,000 Baht) (Leeruttanawisut and Fukushima 2017). In urban areas, only two housing typologies were offered for this price. On the one hand, low-cost 24 m^2 studio apartments and, on the other, 33 m^2 one-room flats. In addition, 60 m^2 two-bedroom two-floor detached houses, row-houses, and twin houses were also offered.

Regarding the development of the mass-housing projects, it took place according to five different scenarios: construction by NHA on public land owned by the agency; construction on other public land with a long-term lease agreement; acquiring completed units from non-performance assets to re-sell them; construction by NHA on private land through public-private partnerships; and through public- private partnerships contracting private developers to construct the housing but with management by NHA (Sangdeeching 2006).

3.1.1 Financial Problems and Limitations in the Implementation Capacity

The very ambitious objective and the scale of the programme proved to be beyond the capacity of NHA and the local authorities (Table 1). By the deadline of the programme in 2007, NHA had provided only 88,129 units, 14.5% of the total. And

Table 1 Baan Eur-Arthorn housing units delivered in the BMR and other provinces

Baan Eur Arthorn housing units delivered				
Year	Bangkok	5 Surrounding Provinces	Other provinces	Whole country
2002	1.417	5.084	11.282	17.783
2003	–	477	–	477
2004	692	836	–	1.528
2005	–	3.113	200	3.313
2006	12.668	25.753	7.612	46.033
2007	2.564	5.212	11.219	18.995
2008	12.395	29.582	18.622	60.599
2009	9.136	40.825	21.884	71.845
2010	10.913	10.341	14.051	35.305
2011	–	4.128	9.890	14.018
2012	417	8.054	2.578	11.049
2013	–	862	743	1.605
Total	50.202	134.267	98.081	282.550
% to whole country	17.77	47.52	34.71	100.00%

Source Author adapted from NHA data 2014

by the end of 2013, six years after the official end of the programme, there were a total of just 282,550 units, not even half of the expected units when the programme started, including the units that were built with the involvement of the private sector. In 2007, the government recognised significant economic losses and financial problems due to the housing surplus. Responding to this situation, and to minimise the losses, the government decided to terminate the programme and requested NHA to reduce the supply target to 300,504 units continuing only with the projects underway. In this way, the programme finalized without reaching the objectives and leaving a bitter taste. Several reasons explain the failure of BEA and the significant housing surplus. It should be noted that not all the units finally produced were sold, and in 2013, 25,969 were still unsold. The first reason that explains the failure is that the typologies offered did not match the expectation of the demand, since most of them were too small for the type of potential buyers. The second is that the estimations done by NHA regarding the income misunderstood the amount that the low-income population could afford, and the GHB did not approve many mortgage applications. And, finally, the increase in the costs and the interest rate after the fifth year resulted in many loaners not being able to pay their monthly instalments (Leeruttanawisut and Fukushima 2017).

3.1.2 The Benefits of the Baan Eur Arthorn Programme

Despite the incapacity to finalize the objective of the Baan Eur Arthorn to build the units promised, the programme did have some positive impacts. In this sense, as mentioned by Leeruttanawisut and Fukushima (2017), the policies implemented to improve the tenure status of the low-income population were very positive. 77.8% of the people who opted for the BEA achieved home or family ownership, which indicates that the programme was successful in achieving this objective. Also, 58.6% of the owners reported improvements in the housing compared to their previous situation, which had an impact on their quality of life. This can be attributed to the provision of community facilities such as day-care, community centres, small stores, playgrounds, and gardens to assist residents and create a good living environment (CODI 2006). However, it should also be noted that, like in many other countries, the production of mass-housing shows too many weaknesses when it comes to solving the problems of the low-income population.

3.2 'Secure Housing'. The Baan Mankong Programme (BMP)

CODI created the Bann Mankong Programme (BMP) following the community-driven legacy initiated by UCDO. The policies implemented in 2003 by the BMP had as their main target the provision of security of tenure to low-income communities by offering long-term lease housing conditions or the possibility of purchasing the property. This objective responded to the identification by the NHA of tenure security as one of the crucial challenges facing informal settlements in Thailand. The essential difference compared to BEA was that, instead of dedicating the efforts to producing housing, BMP worked through a community-driven approach, creating savings groups and empowering communities. By placing low-income residents at the centre, BMP transformed the role of the community, becoming the central actor in the development process. The challenge, in this case, was to support upgrading to allow low-income communities to lead the process and build local partnerships, contributing as part of a city-wide approach. As we will see, the ability to use flexible financial management systems was the key and the foundation for addressing this challenge.

The BMP set a target of improving housing, living conditions and tenure security for 200,000 low-income households, within five years. The programme is currently the primary national reference in upgrading informal settlements with a strong focus on community participation in planning, implementation and funding housing and infrastructure improvements (Bhatkal and Lucci 2015).

3.2.1 Beyond Housing Production. Implementing a City-Wide Program Through Community Networks and Saving Groups

When UCDO started facing a problem of scale and difficulties in supporting individual low-income groups, the creation of community networks was seen as a new mechanism that could help to scale up the development and impact of the program. With this new approach, BMP allowed individual communities to move from isolation into collective strength to develop solutions to the problems they faced. As explained by CODI "besides providing a means of idea-sharing, asset-pooling and mutual support, networks have opened channels for communities to talk to their local development agencies and to undertake collaborative development activities of many sorts."[5]

CODI describes these community networks as platforms for large-scale development, which involve a synergy of learning, experience-sharing, morale-boosting and mutual inspiration, providing low-income communities with confidence via a mechanism that belongs entirely to them. Ultimately, community networks have become the main community-driven development mechanism for CODI, in its work to develop a national-scale development process.

CODI explains the benefits of community networks in seven principles (CODI 2019):

1. Networks offer a platform for dealing with any issue. Network systems draw people together around any subject of importance to them. Links between these networks and CODI's credit and development grant funds, provide communities with a strong capacity to tackle these issues, on their own terms.
2. Networks build collective capabilities at scale. A new culture of "togetherness" is created, providing poor communities with strong capacities to deal with broader and structural issues that would not be possible to be solved by community organisations individually.
3. Networks are learning platforms that work as information channels where, people to learn from each other. Knowhow and successful alternatives are transmitted among community networks.
4. Networks open up community processes in which many community members are involved. The more activities organised by networks, the more people get involved.
5. Networks are internal support systems for people's processes that provide support systems to individual communities, and give Thailand's poor groups confidence, negotiating power, information, and tools to deal with pressing problems in their urban environments.
6. Networks are internal checks and balanced systems that allow equitable community-driven development processes.
7. Networks act as a bridge to the formal system. They demonstrate that there are ways of bridging the gap between the urban poor and the formal systems. This

[5]Ibid., note 2.

has been proved effective due to the strengthened role of community networks in the negotiation processes and the diverse, innovative collaborations established with other stakeholders in the development of projects affecting the urban poor.

Aside from creating community networks, participation is an essential aspect of BMP. Communities have to develop and manage savings groups. These groups will be in charge of planning the upgrading projects and implementing them in collaboration with local authorities. Community groups also set up a cooperative to manage the loan and to lease the land collectively. As indicated by Boonyabancha (2009), "housing is not an end in itself but, rather, the beginning of more community development, in which a group of poor people can live together and can continue to address the real issues of their poverty as a matter of course."

As of today, according to CODI data, there are 77 province-level, five regional-level, and hundreds of active city-level networks. The networks are divided into "area-based", when they focus on specific community locations, and "issue-based" national-level networks, which bring together communities around common issues such as welfare, housing, a common landlord or tenure situation.

3.2.2 Types of the Baan Mankong Upgrading Systems and Land Tenure Options

Although BMP's focus goes beyond physical housing upgrades, it also offers policies and a diversity of planning solutions to solve the physical reconstruction of the community while adapting to residents' needs and priorities as well as the owners'. In this sense, to transform the physical environment of the low-income communities, the programme suggests five alternative systems to upgrade informal settlements.[6]

- On-Site Upgrading

On-site upgrading aims to improve the basic services and the physical environment in existing communities while preserving their location, character and social structures. In general, the neighbourhood layout and plot sizes are not changed, focusing on the upgrading of houses, open spaces and streets. Upgrades seek to improve the physical conditions and quality of life of the communities, as well as incentivising income generation.

- On-Site Reblocking

Reblocking focuses on more fundamental actions regarding the infrastructure and physical environment. This type of on-site upgrading aims to ensure the connectivity and continuity of the existing communities within the city network. It is achieved by introducing adjustments to the street layout and the housing plots to install infrastructure (sewers, drains, walkways and roads) and define the boundaries between private and shared spaces. Although reblocking projects try to avoid impacts on the

[6]CODI 2009. https://en.codi.or.th/download/types-of-development/.

community, they usually require relocating some houses either partially or entirely to improve connectivity and re-align streets to facilitate the layout of infrastructure. If individual dwellings need to be demolished, residents are relocated within the same community through land-readjustment. According to CODI, reblocking is often the system chosen in the cases where communities have negotiated to buy or obtain long-term leases for the land they already occupy.

- On-site Reconstruction

The third system is on-site reconstruction. This system involves the demolition of the existing built-up area and its redevelopment on the same land following a different parcellation. The project is usually developed under a long-term lease or after the people have negotiated to purchase land from the owner. The new security of land tenure provides residents with a powerful incentive to invest in the construction of the new housing. This strategy generates an entirely new physical environment decided by the community and for the community and allows people to continue living in the same place and to remain close to their workplaces and vital support systems. This continuity is a crucial compensation for the expense and difficulty involved in reconstruction.

- Land Sharing

Land sharing is a housing and settlement improvement strategy which benefits both the landowner and the community people living on that land. Land sharing is not a new approach; in the 1980s, the system was already implemented in Bangkok as an alternative to forced evictions (Sheng Yap and De Wandeler 2010). This system requires a process of negotiation and planning, after which an agreement between the community and the landowner is reached. The agreement entails the subdivision of the area into two parts so that both parties involved share the land. A prior planning and discussion stage is needed to establish which part of the land is less commercially viable so that portion is provided to the land-owner, while the rest of the land will be sold or leased to the community for reconstructing their houses. As explained by CODI, the land-sharing process intends to translate conflicting needs and conflicting demands into a compromise that is a "win-win" situation and is acceptable to all parties involved. This trade-off may lead to the community and the landowner getting less area than they had before, either in property or in use. The objective is to transform the low-income residents into legal owners or tenants of their land. At the same time, the landowner recovers a share of land that can be developed for a profit.

- Relocation

Relocation is the only type proposed by CODI that does not deal with on-site upgrading. Two types of relocation can be considered in this system: nearby or not-so-nearby relocation. In both cases, residents are relocated to new sites far from the original location. A 'nearby relocation' is within 5 kilometres from the original settlement, whereas a 'not-so-nearby' relocation is more than 5 km away. Although the objective of this system is to increase land tenure and housing security by providing land use rights, ownership or long-term leases, the challenge is that communities have to cover the cost of reconstructing their new housing and, in some cases, the costs of buying the land. In addition, relocation sites far from the original might also be far from workplaces or schools, which results in additional travel expenses and time, as well as the detachment from the original environment.

The BMP offers a diversity of tenure options that can be used as the result of the negotiation processes between communities and owners. Options such as co- operative land purchase, long-term lease contracts, land swapping, land sharing or long-term user rights are some of the collective legal tenure alternatives that communities can obtain from the process.

3.2.3 Financing the BMP

As part of the BMP's strategy, CODI has promoted community savings as a key strategy for building community-driven development processes, in which the low-income population work out the solutions to their problems. Thus, the emphasis of BMP is on spreading this opportunity by making it easy for people to start savings groups and to get access to loans from CODI's funds.[7]

After the efforts started by UCDO, the savings process received further impetus when the BMP was launched. As mentioned before, to access the Baan Mankong loans, communities are required to form savings groups or cooperatives so that they develop the improvements, the new housing or the infrastructure collectively. They are also required to save 10% of the amount of the loan in a community savings account to qualify for it.

CODI, as the agency implementing BMP, channels the national government funds directly to the savings groups and the cooperatives. According to the agency, from 1992 to 2018, which includes the UCDO period, a total of US$353 million (10,652 million baht) in loans have been approved to benefit 963 community organizations and 405,210 families, most of them (90%) for housing construction and improvements.

It is important to note that as of July 2018, 45% of the loans have been repaid. Taking into consideration that these loans are meant for low-income people who cannot access the formal financial market, the repayment percentage proves that the loans fit well within the objective of enabling the less advantaged to overcome the challenge of financing to improve their living conditions. In other words, the direct transfer of resources has proved to be an excellent solution to empower communities and their self-management capabilities (Boonyabancha 2009).

There are four types of loans provided by CODI[8] (Table 2).

- Loans for housing and land

---

[7]CODI community savings: https://en.codi.or.th/community-finance/community-savings/.
[8]CODI Loans: https://en.codi.or.th/community-finance/codi-loans/.

Table 2 Summary of CODI's loans for the period 1992–2018

CODI credits. Grand totals 1992–July 2018

Type of loan	Interest rate	Terms	Approved loan amount	Outstanding loan amount
Land and housing loans	4%	20 years, repayable monthly	288.20 million US$ (9439.95 million Baht)	131.04 million US$ (4293.14 million Baht)
Holistic development loans	3.5%	10 years maximum	4.90 million US$ (160.06 million Baht)	3.15 million US$ (103.35 million Baht)
Revolving fund loans	6%	3 years maximum, repayable monthly	2.97 million US$ (97.43 million Baht)	443.945 US$ (14.58 million Baht)
Community enterprise loans	4%	10 years maximum, repayment varies according to nature of business	7.48 million US$ (244.99 million Baht)	4.49 million US$ (147.60 million Baht)
Other loans	Varies	Varies	21.66 million US$ (709.73 million Baht)	7.84 million US$ (257.49 million Baht)
Total loans disbursed			325.21 million US$ (10,652.16 million Baht)	146.96 million US$ (4816.16 million Baht)

Source Author adapted from data in CODI website (2019)
1 US$ = 37.76 Baht

Loans are available to cooperatives and savings groups to purchase land and/or build housing. The maximum loan amount per household is US$11,900 (360,000 baht). The conditions of the loans include that they should not exceed 90% of the total house construction cost, the annual interest rate is fixed at 4%, and the maximum repayment term is 20 years.

- Loans for holistic development

The purpose is to enable a community to develop a collective solution to any economic and/or social problems such as refinancing debts, food production, or the creation of economic activities. The community has to submit a proposal identifying the problems and describing how the loan will be used to overcome those problems as a community. In this case, the annual interest rate is 3.5%, and the maximum repayment term is 10 years.

- Loans for community enterprises

The objective of this type of loan is to support the low-income population to set up community-owned businesses to improve the financial capacity of families. The annual interest rate for these types of loans is 4%, and the repayment term should not exceed 10 years.

- Revolving loans

Finally, revolving loans are meant to provide community groups with short-term capital. At the same time, they obtain their funds to use as working capital for community enterprises. In this case, the annual interest rate is 6%, and the repayment term is three years.

These loans have some distinctive characteristics.[9] The first is that they are collective, and borrowers must be part of a group of low-income people who live together in the same community, and who have come together to form a savings group or a cooperative. In either case, they need to designate leaders and a committee who will act on behalf of the group members managing the legal documents and loan contracts between the group and CODI. The members of the savings groups and cooperatives have to agree on their terms and objectives, deciding how to save and how savings can be used.

The second characteristic is that the loan process involves community people from the very beginning. The proposal prepared by the community group is studied by city and regional committees, which include the participation of local government and other representatives from other community groups. Once accepted, the proposal is submitted to CODI's national loan committee, which includes CODI representatives, government banks, and community leaders. If the project proposal receives an endorsement from other communities, it provides BMP with the comprehensive city-wide approach needed to have an impact beyond the implementation of small-scale projects.

The third characteristic is that the loans are designed to accommodate the income and financial realities of low-income people. In this sense, CODI applies a set of measures that adapt to the population with fewer resources, such as interest rates lower than the ones offered by banks; a fixed rated throughout the loan period, different types of loan repayment plans and terms; and fixed monthly instalments. Another characteristic is that community organizations are allowed to add a margin of 2% or 3% to the loan's interest rate to cover management expenses and a savings fund for late loan repayments. Finally, the designated community leaders and community committee members provide a personal guarantee on the loan. This requirement implies that community representatives are fully committed to the project, as they have a personal liability as well.

3.2.4 BMP in Numbers. Impact

As we have indicated, since its inception, CODI's primary focus has been to support and empower the low-income population living in insecure housing conditions. According to the Asian Coalition of Housing Rights (ACHR) (2017), there were

[9] Ibid., note 8.

Table 3 CODI housing performance

CODI housing performance including all categories As of May 2018			
Housing program	Projects approved	Number of households	Number of cities, districts or wards involved
Urban housing (Baan Mankong urban)	1.042	103.583	343
Rural housing (Baan Mankong rural)	93	6.824	71
Canal housing	38	5.023	7
Homeless housing	7	604	3
Sub total	1.180	116.034	
Sufficient housing program for the poorest (Baan Por Pieng)	3.706	39.194	2.391
Temporary housing for communities affected by disasters, evictions	66	2.723	66
Total	4.952	157.951	

Source Author adapted from data in CODI website (2019)

1903 savings groups in 345 cities in Thailand, with about 850,000 members and total savings of some 3.2 billion baht (US$102 million).

As of May 2019, the programme has responded to the problems of low-income communities by approving a total of 4,952 housing projects, benefiting and providing tenure security to 157,951 households, which amounts to an impact on approximately 800,000 people. Although CODI's efforts have not eradicated poverty, it should be noted that the housing approach not only aimed to solve shelter and security issues but intended to have a more substantial impact. In this sense, the programme achieved the objective of being "one of the most powerful tools for poverty reduction, and for creating a more balanced, participatory and equitable world".[10] Housing development is, therefore, considered a strategic approach that brings people together, creates larger communities, and provides the physical foundations for people's social support systems in cities as well as in rural communities (Table 3).[11]

Although the figures show a significant impact from the programme, and this has managed to improve the quality of life of a substantial number of low-income people, the benefits have reached only slightly more than half of the original target of 300,000 households. This is due to the complex collaborative process that the program implies, which affects the speed of the implementation process. Also, it should be

[10]Community Organization Development Institute (CODI) https://en.codi.or.th/baan-mankong-housing/land-reform-porgram/.

[11]Ibid., note 2.

mentioned that the qualifying prerequisites to establish community-saving groups and cooperatives can be rigid and demanding and may exclude from the process those who do not have repayment capacity or the skills to manage funds—the most vulnerable within the vulnerable. Thus, there is a need to take into consideration the heterogeneity and diversity of the low-income groups and their subgroups. It is important to find systems to adapt the housing solutions to these different groups as well.

4 Evaluating the BMP Impacts

In 2003, as part of BMP, CODI started working in 10 pilot communities to address land and housing problems affecting the low-income population. The last part of this article evaluates the BMP impact in two of these pioneer projects of on-site upgrading in Bangkok, which were selected as pilot experiences. By examining the on-site upgrading cases of the neighbourhoods of Charoenchai Nimitmai and Ruam Samakee, it is possible to observe whether the community residents have benefited from the upgrading process and in what ways. The specific objective of the survey is to analyse the improvements in the living conditions of the residents, comparing their previous conditions and dwelling places to their new residences. This is done by examining the economic, social, and physical impacts through aspects such as tenure status, who the final beneficiaries were, housing ownership, and the housing and living conditions.

A structured survey with 92 questions was carried out in 2018 in both sites, using a questionnaire administered through face-to-face interviews. The questionnaire was divided into four parts regarding the land and housing tenure status and the characteristics of the built environment, the economic status, the residents' income and savings status, and the level of satisfaction regarding the improvements. Each part of the questionnaire has two parts to analyse the status before the upgrading and the present conditions. For the two studied neighbourhoods, the survey included a sample of at least 50% of the households, which were selected randomly representing a diversity of genders, ages, and roles within the community. Surveys were conducted by the author and a team formed by city officials from local government bodies, community representatives, a research assistant, and a group of students from the Faculty of Architecture, Chulalongkorn University. The presence of local officials, community representatives and students helped to create a productive dialogue with the residents. Students also helped to translate the surveys into English, which was essential to conduct further analysis of the gathered information. A review of these BMP pioneer experiences can offer valuable lessons for other informal communities and cities undergoing similar processes.

4.1 *Ruam Samakee. A Land-Sharing Scheme*

Ruam Samakkee occupies 0.9 ha of land of the Crown Property Bureau (CPB) in Ramkhamhaeng Soi 39 area in Bangkok. Originally, the area was an interstitial urban space that started to be occupied by rural migrants in the early 1990s. As the location was an access point to many job opportunities, the informal settlement grew rapidly through social connections becoming an dense and overpopulated community of 124 families.

In 2003, after the adoption of the BMP, the community of Ruam Samakee was selected as one of the 10 pilot projects and 1525 units that became the first models for other communities for future housing development. In May 2003, the people began working with young architects from CODI to develop the new layout plan for the neighbourhood. In the three following months, they demolished all the old houses, raised the level of the land to prevent flooding and laid the new streets and infrastructure. By the end of 2004, the construction of 82 housing units was completed, and in 2008 the development of the 124 units was completed, which increased the perception of security of tenure among residents (UN-Habitat 2009a, b; Archer 2010) (Fig. 2).

4.1.1 Land and Tenure Status

The land sharing upgrading process of Ruam Samakee took a long time. It officially started in 1998, before BMP, when the CPB planned to develop the area and decided to lease the land to a private developer who would deal with the eviction of the existing slums. The community rapidly started to organize and negotiate with CPB to regularize their tenure status. After registering as a cooperative and setting up a savings group, the community negotiated a 30 year collective lease on the land they were occupying. The agreement required the community to return a plot of 0.16 ha to the CPB while they kept the rest of the area at a monthly rental cost of

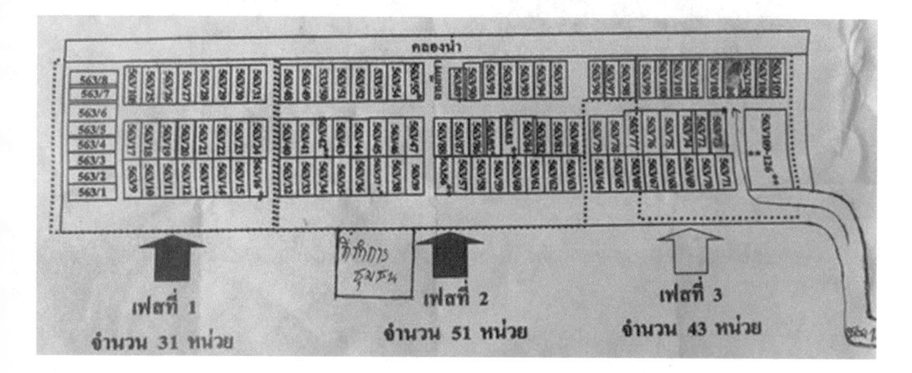

Fig. 2 Land division. *Source* Author photo from community map

approximately US$6 (180 Baht) per household to rebuild their houses (UN-Habitat 2009a, b). The deal, which was very favourable for the community, was the first long-term lease provided by CPB to informal settlers, since on previous occasions the land leases granted were only for one year. All these benefits provided the community with security of tenure to further invest in the demolition and reconstruction of their houses (Sim Junhui 2014).

4.1.2 The Physical Environment of the Upgraded Neighbourhood

The new layout consisted of a grid plan formed by a 6-m wide and 250-m long street as the main spatial space for the community, also accommodating a small public space and a community centre. The new housing is arranged in three rows, two flanking the main street and one facing the exterior road. The land is divided into 108 plots, each measuring approximately 4.5 m wide and 11 m deep, and one larger plot for apartments (Fig. 3).

With support from architects from CODI, the community developed three housing types to adapt to the different family sizes and affordability. The designed typologies included semi-attached twin houses, row houses and apartments. The decision of including apartments was taken due to the limited area and the necessity to accommodate all the settlers. This created an issue of preference, because the majority of the residents preferred to live in houses instead of apartments. The situation was solved by establishing the order of allocation starting from those residents who had been on the site for a longer time. Once all the houses were distributed, the remaining residents were given the option of either living in rental apartments or leaving the community receiving economic compensation. As explained by the residents during the surveys, since the long-term cost of living in rental apartments was more than living in houses, those who could afford it decided to stay in the apartments, but others chose to leave the community.

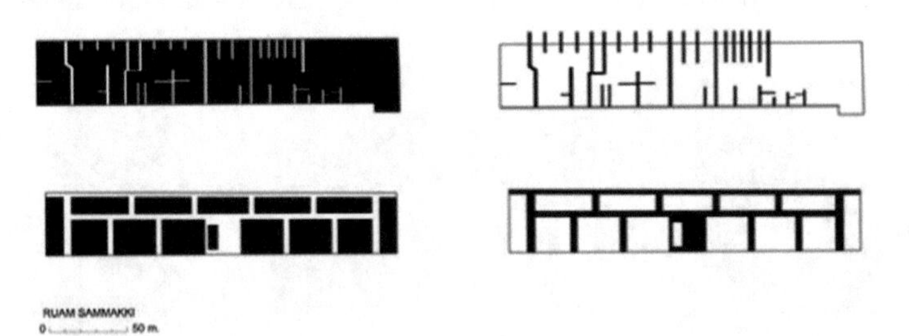

Fig. 3 Analysis of the built-up space and street layout before and after upgrading. *Source* Created by Author

4.1.3 Evaluating Ruam Samakee

- Secure Legal Tenure

One of the main focuses of the survey studied is to examine whether the people who were involved in the upgrading process are still the residents. This is an important indicator that shows if the initiative has succeeded in providing the low-income population with a way to settle permanently—which was the objective of the BMP. In this sense, we observe that only 68.3% of the residents are the original owners. This low percentage indicates that a significant number of the original community decided to leave the area and go to other places

Regarding the improvements in their tenure status, we observe that 90% indicated that they do not own the land, but they own the house and have a title. Compared to the status before the upgrading, we observe that most of the neighbours, 40%, were renting, and 16.7% were already owners of their houses. Significantly, 26.7% of the residents had unsettled ownership or they just held it for free. Also, 68.3% indicated that the current legal housing tenure had provided them with a sense of security. In this sense, even though CPB still owns the land, the land-lease agreement indicates a very high level in the sense of tenure security, which was missing previously.

Aside from this, it is interesting to note that 80% of those surveyed indicated that they have not always lived in the same house in the neighbourhood and that 66.7% currently own more than one property, which they rent as an extra income. This indicates the attachment of some of the neighbours to the area and how improvements have allowed them to progress economically. However, it also suggests that some of the housing properties have lost their initial social objective of providing housing for the original settlers and have been transformed into part of their income.

- Improved Built Environment

71.7% of the residents in Ruam Samakee are satisfied with their current house, stating that the quality has improved significantly with the upgrading. Before upgrading, 30% of the residents were living in shanty houses of one (50%) or two (25%) rooms, and 15% in rented rooms, whereas, currently, all the residents live in proper houses or apartments of two (51.7%) or three (23.3%) rooms. As a conclusion, 93.3% expressed their preference for the current house compared to the previous house, which indicates that the project succeeded in providing a better living environment for the low-income population both in quality and size adapted to their needs. This is important, because the assessment showed that about 37% of the community members used their houses as a means for income generation, such as having the front part of the house as a grocery shop, food stall, car workshop or laundry (UN-Habitat 2009a, b).

Regarding the basic services, as mentioned, the upgrading project implemented a new drainage system and septic tanks, water supply, electricity, as well as a community centre and a community park. This is reflected in the survey, in which all the residents (100%) expressed satisfaction about the new services provided. However, a significant number of residents mentioned that access to education and healthcare needs to be improved, even though these two items were not part of the upgrading project. Despite this, it is also an indicator of the improvements, because residents have started asking for other services and not just the basic needs.

- Economic Improvement

The questions related to economic aspects show improvements for most of the families. Concerning the household income, 56.1% have seen an increase after the upgrading project. At the same time, 15.8% have the same amount, and 15.8% used to have more income. This allows to 55.9% of the households to have monthly savings compared to only 38.2% in the past, which shows that the new situation has improved the economic status of the families significantly. However, we observe that the majority (71.9%) do not invest their savings and use them for emergencies and other family needs, although investment has increased by 13.8 points.

However, although income increased, expenses did as well. Residents indicate that, in the past, expenses were lower before in 68.6% of cases. Moreover, we observe that to reach this improved household income and to cover expenses, more household members have to work. Thus, while the number of households where one or two members earn an income is similar to the past, households that need three and four members to earn income have increased from 14.6 to 32.2%, more than double. However, this also shows a positive aspect, because more family members can get jobs due to general household improvements. It is also notable that the household income has not fluctuated in recent years (80.3%), which shows significant stability in jobs and family earnings.

4.1.4 Conclusions About the Overall Perception and Satisfaction in Ruam Samakee

76.7% of the residents expressed satisfaction living in the upgraded Ruam Samakkee, and only 3.3% said they are unsatisfied. Regarding the evaluation of services, there is a common agreement that they are much better than before the upgrading. Also, the perception has changed regarding the way residents see themselves. In this sense, while in the past 38.3% felt like squatters, currently less than 2% have this feeling (Fig. 4).

Finally, in terms of safety, 81.7% of the respondents indicate that there is adequate physical safety in the community concerning crime, drugs and violence. More than half of the residents (63.3%) reported that they feel safer in the upgraded community compared to the previous situation. However, still, a considerable amount (30%) of residents express concerns over safety and feel that there has not been much improvement in the safety of the community as compared to the previous settlement.

To conclude, although CODI's approach claims to be community-driven, according to the Ruam Samakee residents the community participation in the design process of the new neighbourhood was limited to consultation and consensus by voting on their preference from options provided. This fact suggests that there was minimal dialogue in the decision-making and that the process of planning and design

Fig. 4 Current status of Ruam Samakee community. *Source* Author

was done more from a top-down perspective than from a collective or bottom-up approach, although the construction phase incorporated more participation from the residents in the building works. However, it is also important to note that Ruam Samakee is one of the first experiences, and the collaborative and community- driven system was still developing.

The residents of Ruam Samakee are determined to keep upgrading their community and their standard of living. The environment of the settlement is extremely positive and inviting, which has helped the community to fit into the urban fabric of the city easily. Although there are some aspects to be improved, the advantage of the BMP programme providing on-site upgrading has proven to be positive to keep the community in place and from the socio-economic and environmental perspectives.

4.2 Charoenchai Nimitmai. An On-Site Reblocking Scheme

The community of Charoenchai Nimitmai is one of the informal settlements located between the Prem Prachakon drainage canal, the railway tracks, and the Sirat expressway in Bangkok's Chatuchak district. Originally, 41 families were living on the 0.7 ha, who had been renting the land from a private landowner for over 50 years. In 1995, UCDO surveyed the community as part of its national survey of urban slums and encouraged the community to form a savings group. As explained by UN-Habitat (2009a, b), five women began the first group, and three years later, in 1998, all 41 households were saving and generated a total of US$99,500 (3 million Baht). That year, threatened with eviction when the landowner put the land up for sale, the community started negotiations to buy the land for around US$497,500 (15 million Baht), a fifth of its market value (market price was established at 87 million Baht). After creating a cooperative, and with the aid of a CODI loan of 18 million Baht at a 2% interest rate, they acquired the land in 2000 (Sim Junhui 2014; CODI (Community Organization Development Institute) 2006), using the savings as a guarantee for the loan from CODI.

4.2.1 Land and Tenure Status

Like Ruam Samakee, Charoenchai Nimitmai was selected by CODI to be one of the pilot projects in the BMP. Once the community managed to buy the land, they started an on-site reblocking project increasing the number of plots to bring down the land and construction costs per family. With the support from CODI architects, accountants, and social facilitators (Fig. 5), the community set up a new spatial site plan with a total of 89 plots of varying sizes, incorporating a community centre and the 48 new parcels to accommodate vulnerable families squatting nearby, that would help to bring down costs. In order to fit the new street layout and plots, 15 houses had to be demolished and relocated within the site (Fig. 6). Despite the impact of the proposal, the planning process was collective and very inclusive, and according to the surveys, residents responded that the majority of the community members participated in the design of the layout during a six-month period, and underwent

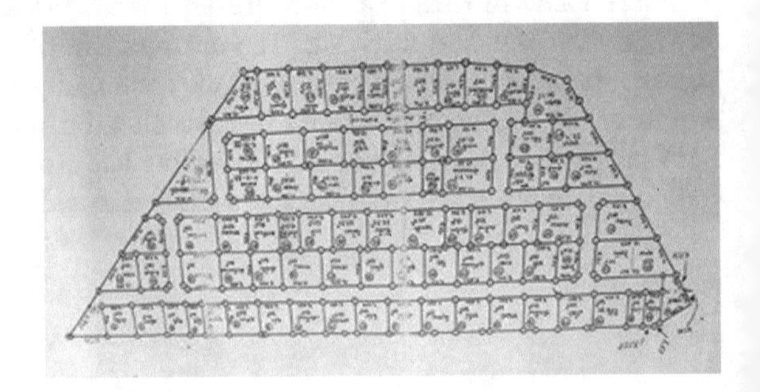

Fig. 5 Land division. *Source* Author photo from community map

multiple iterations of planning, going through 22 meetings, 18 different layout plans as a part of the process (Sim Junhui 2014; UN-Habitat 2009a, b). As a result of the process and the planning, the community established collective ownership of the land.

4.2.2 The Physical Environment of the Upgraded Neighbourhood

The proposed spatial site plan consists of a grid formed by three parallel streets, each one 3.5 meters wide. Along the streets, plots from 40 to 100 m^2 are organized linearly (Fig. 7). The community replaced their old wooden walkways over water and mud with concrete walkways wide enough for cars to pass through (Groen 2011). The final layout of the community compared to the previous status before the upgrading shows a very significant decrease in the percentage of outdoor space. This factor is due to the substantial increase in the housing density (219%), due to the occupation of the existing open space in-between the houses of the original community. While this decrease also leads to a reduction in absolute terms of the ratio of outdoor space per person, it is interesting to note that, when comparing only the street space, discounting the space in between buildings, the ratio of outdoor space per person increased by 357%, from 1.4 to 5, which is far from negligible.

4.2.3 Evaluating Charoenchai Nimitmai

- Secure Legal Tenure

In the case of Charoenchai Nimitmai, we observe that 80% of the residents are still the original owners. In contrast to Ruam Samakee, this indicates a stronger self-containment capacity and an achievement in providing long-term housing tenure to the original low-income residents. Also, in comparison to Ruam Samakee, in this case, 83.3% indicated that they have always lived in the same house in the neighbourhood since they moved in, again proving the self-containment and attachment capacity of this project. Only 20% of those surveyed indicated that they own another property, which is an indicator of the lower-income level of the residents in Charoenchai Nimitmai compared to the residents in Samakee.

Regarding the improvements in their tenure status, we observe that 96.7% indicate that they own the house and land as part of the community, which still follows the agreement with CODI. Compared to the tenure status before upgrading, we observe that, as in the previous case, most of the residents were renting (43.3%). However, it is noticeable that 23.3% indicate that they were already owners of the house, which includes the relocated people from nearby neighbourhoods. However, 90% of the previous owners say that they sold the property. Finally, concerning the sense of tenure security, 93.3% indicate that it has increased because of the ownership of land and house provided by the upgrading project. This very high percentage indicates

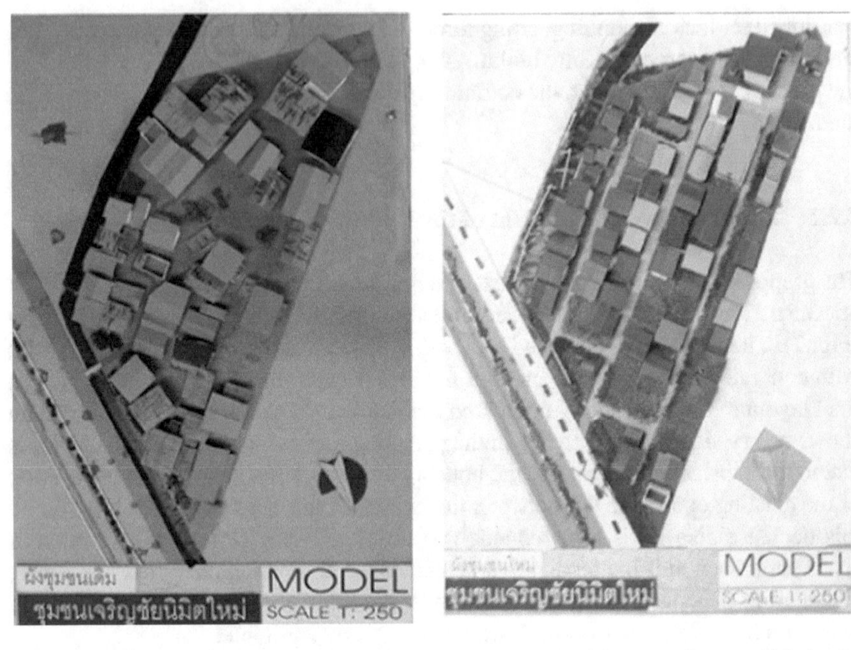

Fig. 6 Models showing before and after layouts. *Source* Author photo from models in the community

the success of this reblocking scheme, much higher compared to the case of Ruam Samakee.

- Improved Built Environment

Regarding the built environment, 90% of the residents in Charoenchai Nimitmai are satisfied with their current house, a significantly higher proportion than the previous case and again an indicator of the attachment to the area. This is also more remarkable when we cross the information with the type of housing they had previous to the upgrading. 83.3% indicate that they were already living in houses of mostly two or three rooms (50 and 23.3% respectively), and only 6.7% used to live in shanty houses. To the status after the upgrading, all the residents (100%) indicate that received a house—40% two rooms and 33.3% three rooms. Although most of the neighbours already had a house before the upgrading, 86.7% expressed their preference for the current house compared to the previous house. This means that, although the housing standards before upgrading were quite high, much better than in Ruam Samakee, the residents in Charoenchai still have a higher degree of satisfaction and sense of improvement, a good indicator of the success of the upgrading project in providing a better living environment for low-income residents. Concerning the basic services, the upgrading project included the provision of electricity and electricity meters, septic tanks for waste disposal, and also the local government and the BMA agreed to collect the solid waste from the area (Groen 2011). These improvements are

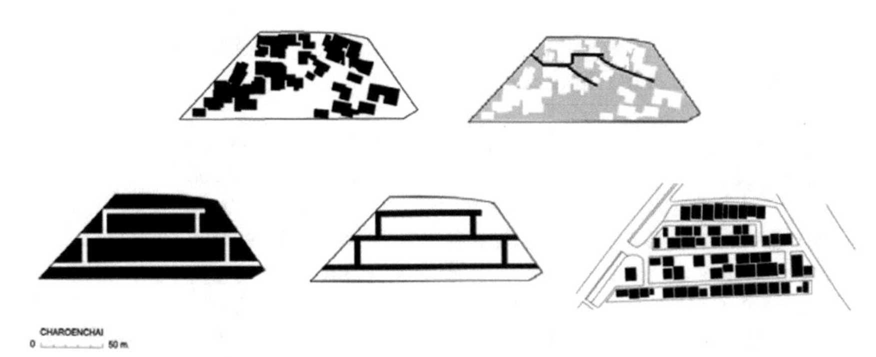

Fig. 7 Analysis of the built-up space and street layout before and after upgrading. *Source* Created by the author

reflected in the survey, in which all the residents (100%) expressed satisfaction with the water, electricity, sanitation, drainage, garbage collection, and street infrastructure. However, as in Ruam Samakee, more than 20% expressed their dissatisfaction regarding education and health services, which are described as insufficient.

- Economic Improvement

Concerning income, the survey indicates that it has increased in 69% of the households after the upgrading project. At the same time, 10.3% have the same amount and also 10.3% used to have more income. Therefore, the residents in Charoenchai seem to have improved economically more than in Samakee. This allows 70% of the households to have monthly savings, a higher percentage than in Samakee, and also higher compared to the 35.7% who were able to save in the past in Charoenchai. However, we also observe that in Charoenchai, despite more people investing their savings than in Samakee, the majority (65%) still prefer not to invest their savings, although in this case investment has increased by 15 points. Regarding expenses, they also increased, like in Samakee. Residents indicate that, in the past, expenses were lower than currently—in this case, in 85.7% of the households. In this sense, like in Samakee, the number of economic sources had to increase and, although the number of households with one or two members earning income has remained stable, households where three salaries are needed have doubled, from 10.3 to 20%, and those where four income sources are required have increased 45% (from 6.9 to 10%). Although this is a good indicator of the access to the formal job market, it also indicates the economic demands of the upgraded situation. However, similarly to Ruam Samakee, it is good to observe that the household income in Charoenchai Nimitmai has not fluctuated in recent year (72.4%), which indicates stability for families facing such demanding scenarios.

4.2.4 Conclusions About the Overall Perception and Satisfaction in Charoenchai Nimitmai

In Charoenchai, 86.7% of the residents expressed satisfaction about living in the upgraded neighbourhood, and none of them felt unsatisfied. Perception of their status has also changed, although not as significantly as in Samakee. In this sense, while in the past 23.3% felt like squatters, currently 6.7% have this feeling, which is a good indicator of how the project has improved their quality of life.

Finally, regarding the perception of safety, 93.3% of the respondents indicate that the new upgraded neighbourhood has adequate physical safety in the community with regard to crime, drugs and violence. However, only 53.3% reported that they feel safer in the upgraded community compared to the previous situation, and 36.7% said the safety level was the same.

To conclude, it can be said that the success of the on-site upgrading project of Charoenchai Nimitmai is thanks to the combination of different aspects: the support from agencies; an active and strong group of community leaders; a very intensive iterative discussion process, in which all the community members took an active part; and the honesty, transparency and enthusiastic work done by people (Fig. 8).

5 Concluding Remarks

As stated by Bhatkal and Lucci (2015), the Baan Mankong Programme constitutes one of the few programmes worldwide that address the upgrading of informal settlements at the national level using a community-driven approach. In this sense, it can be said that the programme has not only promoted the physical improvement of communities but has also strengthened the social cohesion and social status of low-income communities due to the relationship that exists between the increase in tenure security and the investments in housing improvements.

More specifically, the BMP, as a supportive, integrated, inclusive, and somewhat bottom-up initiative, has proven to be significantly successful in achieving one of the main objectives, land and housing tenure security. This success has come about because, rather than just being a supply-driven approach by government agencies, which is the conventional approach of most housing programs, the BMP is a flexible and demand-driven approach that puts communities at the centre of the upgrading process and supports tailored responses to their needs and priorities (Boonyabancha 2005).

In this sense, communities under the BMP take an active and committed role in the upgrading process that goes beyond being mere beneficiaries of a publicly subsidized program. In the community-driven approach, collective contributions from savings groups and networks play an essential role in controlling and managing funds, as well as in the designing and building of the physical environment, which allows funding to be more efficiently used and targeted to their strategies and interests. The creation of community groups and networks has played an essential role in

Fig. 8 Current status of Charoenchai Nimitmai community. *Source* Photos by Author

empowering and building capacity to address residents' specific needs, as well as in stimulating confidence and social changes among the low-income population.

However, there is still room for improvement, and a focus on policies to prevent the formation of informal settlements should also be included. Usually, upgrading policies and programs react to current scenarios; however, it is not common to see a vision that plans to mitigate the effects of informality. Similarly, the Baan Mankong Program focuses on existing informal settlements but does not focus on the incoming population that will generate those settlements.

Policies and programmes to produce affordable mass-housing to absorb the incoming low-income population might help to prevent the formation of new informal settlements. However, in many cases like the BEA, they have proved to be insufficient and limited, and when they deal with existing informal settlements, they are not the appropriate solution. Therefore, there is a need to establish mechanisms to deal with the integration of both phenomena—the existing informal settlements and the incoming population. Effective governance, urban planning and policies with a comprehensive vision that integrates housing can generate the framework to simultaneously prevent the formation of informal settlements in addition to their upgrading. With this purpose, urban planning departments should be provided with independence, policy tools and management and implementation capacity to provide such a framework.

In this sense, on-site upgrading projects cannot be the final objective in themselves. They should go beyond just scaling-up small initiatives and the "one project at a time" approach. On-site upgrading needs to be part of a city-wide strategy and comprehensive plan that is collaborative, inclusive and integrates the poor. Conventional planning and policies are not devised with this aim, and usually they do not incorporate the needs of this sector of the population. Therefore, with regard to the low-income population, there is a need for an inclusive planning process determined and controlled by low-income people that also engages and incorporates the different stakeholders and groups affected by the development of these pro-poor strategies and city-wide plans. With this approach, low-income communities will become integrated and legitimate parts of the city development process. Thus, the cooperation and partnerships between the different actors is crucial to the success of the informal settlement upgrading programs, in which public agencies are no longer the only city-makers delivering for beneficiaries.

To conclude, good governance for informal settlement upgrading should be adaptive to respond to the people's needs by using their initiatives as a framework; it should be flexible and diverse, offering a variety of means to achieve the main objectives; it should provide assistance, resources and infrastructure to the low-income population, building capacity and promoting self-sufficiency; and, lastly, it should have the potential to be scaled up to serve as a city-wide model for similar situations.

References

Archer D (2010) Empowering the urban poor through community-based slum upgrading: the case of Bangkok, Thailand. In: 46th ISOCARP Congress 2010
Asian Coalition for Housing Rights. ACHR (2015) 215 cities in Asia: fifth yearly report of the Asian coalition for community action. Asian Coalition for Housing Rights, Bangkok
Asian Coalition for Housing Rights. ACHR (2017) Community finance in five Asian countries. Asian Coalition for Housing Rights, Bangkok
Bangkok Metropolitan Authority. (2001) 9 years of Community Development Department. Airborne Print, Bangkok
Bhatkal TLucci P (2015) Community-driven development in the slums: Thailand's experience. Overseas Development Institute
Boonyabancha S (1999) Citizen networks to address urban poverty: experiences of urban community development office. Thailand, Unpublished
Boonyabancha S (2000) Providing housing for the poor: a new approach to achieving comprehensive housing development by urban poor communities and the city, as gathered from experiences in Thailand, UCDO Thailand
Boonyabancha S (2002) Community as subject, not object: using a Community Development Fund as a flexible tool to support community initiatives and to help people build their own social well-being experiences from CODI, in Thailand. Paper presented to the international symposium on social well-being and development: toward a policy science in support of community initiatives
Boonyabancha S (2003) A decade of change: from the urban Community Development Office (UCDO) to the Community Organizations Development Institute (CODI) in Thailand. Human Settlements Working Paper Series Poverty Reduction in Urban Areas. No. 12. International Institute for Environment and Development, London
Boonyabancha S (2004) The Urban community development office: increasing community options through a national government development programme in Thailand. In: Diana M, David S (eds) Empowering Squatter Citizen. Civil Society and Urban Poverty Reduction, Earthscan, London, Local Government, pp 25–53
Boonyabancha S (2005) Baan Mankong; going to scale with 'slum' and squatter upgrading in Thailand, Environ Urbanization 17(1)21–46
Boonyabancha S (2009) Community Development Fund in Thailand: a tool for poverty reduction and affordable housing. UN-HABITAT
Boonyabancha S, Kerr T (2018) Making people the subject: community-managed finance systems in five asian countries. In: Environment & urbanization, 2018, vol. 30(1). International Institute for Environment and Development (IIED), pp 15–34
Carracedo O (2013) Regenerating neighbourhoods through identity and community places: planning and developing strategies for resilient informal settlements. In: Conference proceedings. 7th conference of the international forum on urbanism (IFoU). Tainan, Taiwan
Carracedo, O. (2014) Shaping informality. The role of street-based strategies in revitalizing informal and low-income areas. In: 7th International urban design conference. Adelaide, Australia
Carracedo O, Noguer N (2013) Upgrading suburbs in the Latin American context. A management and transformation review of the slums. In: Conference proceedings. 25th Conference of the European Network of Housing Research, Tarragona
Carracedo O, Noguer N (2013) Retrofitting suburbia and informal slums in the Latin American context. Recovering neighbourhoods in extreme poverty conditions with tactical urbanism and new urbanism principles. In: 21st conference of new urbanism, Salt Lake City
Castanas N, Yamtree PK, Sonthichai YB, Batreau Q (2016) Leave no one behind: community-driven urban development in Thailand. IIED Working Paper. IIED, London
CODI (Community Organization Development Institute) (2005) Baan Mankong: an update on city-wide upgrading in Thailand, Bangkok. Community Organization Development Institute

CODI (Community Organization Development Institute) (2006) Urban development towards sustainable cities and housing for the urban poor in Thailand. http://www.codi.or.th/downloads/english/Paper/Urban_Poor_in_Thailand_062006.pdf. Accessed on 15 Oct 2019

CODI (Community Organization Development Institute) (2009) Workshop on shelter security and social protection for the urban poor and the migrants in Asia. Unpublished. Presented: Ahmedabad, India, 11–13 Feb 2009

CODI (Community Organization Development Institute) (2019) The key partners in all of CODI's work: community networks. https://en.codi.or.th/community-network-activitys/why-network/. Accessed on 15 Oct 2019

Glaeser EL (2011) Triumph of the city. Penguin Press, New York

Grindle M (2004) Good enough governance: poverty reduction and reform in developing countries. Gov Int J Policy Admin Inst 17(4)

Groen A (2011) Building towards safety: tenure security & the improvement of living conditions in Bangkok Slums. University of Amsterdam. Faculty of Social and Behavioural Sciences. Department of Urban Planning, Human Geography and International Development Studies. Unpublished

Leeruttanawisut K, Fukushima S (2017) An evaluation of the Baan Eua Arthorn housing program in thailand: risks of a populist housing policy. Urban Reg Plann Rev 4:2017

Na Thalang W (2007) Special interview: historical perspective. GH Bank Hous J 1(1):2–6. National Housing Authority. Retrieved 23 Jan 2017, from https://www.nha.co.th/view/2/HOME#OURPROJECT. Accessed on 15 Oct 2019

National Economic and Social Development Board (NESCB) (2016) The national economic and social development plan. Office of the Prime Minister. Bangkok, Thailand. https://www.nesdb.go.th/nesdb_en/ewt_w3c/main.php?filename=develop_issue. Accessed on 15 Oct 2019

National Housing Authority (NHA). (2002) Housing development program for slum and urban poor for national economic and social development plan IX (2002–2006). National Housing Authority

National Housing Authority (NHA) (2005) Annual Report, 2004. National Housing Authority, Bangkok

National Housing Authority (NHA) (2014) Baan Eua Arthorn's project completion. Bangkok. National Housing Authority. Unpublished

Obeng-Odoom F (2013) Governance for pro-poor urban development: lessons from Ghana. Routledge, London

Ouyyanont P (2008) The crown property bureau in Thailand and the crisis of 1997. J Contemp Asia 38(1):166–189

Pornchokchai S (2003) Global Report on Human Settlements 2003. City Report, Bangkok

Pornchokchai S (2008) Housing finance mechanism in Thailand. UN-Habitat, Nairobi

Prachuabmoh K (2007) An integrated national housing development strategy—key to a sustainable Thai economy. GH Bank Hous J 1(1):9–14

Sangdeeching P (2006) An implementation of low-income housing policy: a case study of Baan Eur-Arthorn project and Baan BOI project. Unpublished. Master thesis. Chulalongkorn University, Bangkok

Senanuch P (2004) An investigation into the Policy for Urban Poverty Alleviation in Thailand through the Study of Urban Slum Communities. University of Sydney, Sydney

Sheng Yap K, De Wandeler K (2010) Self-help housing in Bangkok. Habitat International 34(3):332–341

Sim Junhui D (2014) Autonomy of a community: subjectivity of participation in the Baan Mankong Housing Programme. Unpublished. March Dissertation. National University of Singapore

Teokul W (1999) Social development in Thailand: past, present and future roles of the public sector. ASEAN Econ Bull 16(3). Social sectors in South-East Asia: role of the state (December 1999). Published by: ISEAS—Yusof Ishak Institute, pp 360–372

The World Bank (2014) Thailand economic monitor. Bangkok

UN-Habitat (2003) The challenge of slums: global report on human settlements. United Nations Human Settlements Programme, Nairobi

UN-HABITAT (2009) Community development fund in Thailand: a tool for poverty reduction and affordable housing
UN-HABITAT (2009) Slum upgrading facility: exchange visit to the Community Organisations Development Institute in Thailand
Unpublished paper for the UNCHS (2019) https://www.thaiappraisal.org/pdfNew/HABITAT1new.pdf. Accessed on 15 Oct 2019
Viratkapan V, Perera R (2006) Slum relocation projects in bangkok: what has contributed to their success or failure? Habitat Int 30(1):157–174
Yap KS (1992) The slums of Bangkok. In: Yap KS (ed) Low-income housing in Bangkok: a review of some housing sub-markets. Asian Institute of Technology, Bangkok, pp 31–48

Printed in the United States
by Baker & Taylor Publisher Services